THE MAKING OF AN AMERICAN

★

APPALACHIAN ECHOES

Richard D. Starnes, Nonfiction Editor

THE MAKING OF AN AMERICAN

The Autobiography of a Hungarian Immigrant, Appalachian Entrepreneur, and OSS Officer

MARTIN HIMLER

Edited by Cathy Cassady Corbin,
with an Introduction by Doug Cantrell,
and a Foreword by Charles Fenyvesi

THE UNIVERSITY OF TENNESSEE PRESS | KNOXVILLE

The Appalachian Echoes series is dedicated to reviving and contextualizing classic books about Appalachia for a new generation of readers. By making available a wide spectrum of works—from fiction to nonfiction, from folklife and letters to history, sociology, politics, religion, and biography—the series seeks to reveal the diversity that has always characterized Appalachian writing, a diversity that promises to confront and challenge long-held stereotypes about the region.

Introduction copyright © 2018 by Doug Cantrell.
Editing and annotations copyright © 2018 by Cathy Cassady Corbin.
Foreword copyright © 2018 by Charles Fenyvesi.

All Rights Reserved.
Manufactured in the United States of America.
Paperback: 1st printing, 2018; 2nd printing, 2023.

Library of Congress Cataloging-in-Publication Data
Names: Himler, Martin, 1888–1961, author. | Corbin, Cathy Cassady, editor. | Cantrell, Doug. | Fenyvesi, Charles, 1937–
Title: The Making of an American: The Autobiography of a Hungarian Immigrant, Appalachian Entrepreneur, and OSS Officer / edited by Cathy Cassady Corbin; introduction by Doug Cantrell; foreword by Charles Fenyvesi.
Description: First edition. | Knoxville: University of Tennessee Press, 2018. | Series: Appalachian echoes non-fiction | Includes bibliographical references and index. |
Identifiers: LCCN 2018007557 (print) | LCCN 2018030474 (ebook) | ISBN 9781621904526 (pdf) | ISBN 9781621904519 (paperback)
Subjects: LCSH: Himler, Martin, 1889-1961. | Hungarian Americans—Biography. | Himler Coal Company—Biography. | Coal miners—United States—Biography. | Newspaper editors—United States—Biography. | Hungarian newspapers—United States—History—20th century. | World War, 1939-1945—Personal narratives, American. | War criminals—Hungary. | Intelligence officers—United States—Biography. | United States. Office of Strategic Services—Officials and employees—Biography. | BISAC: BIOGRAPHY & AUTOBIOGRAPHY / Personal Memoirs. | HISTORY / United States / State & Local / South (AL, AR, FL, GA, KY, LA, MS, NC, SC, TN, VA, WV). | HISTORY / United States / 20th Century. | BUSINESS & ECONOMICS / Industries / Energy Industries. | HISTORY / Military / World War II. | HISTORY / Europe / Austria & Hungary. | HISTORY / Holocaust.
Classification: LCC E184.H95 (ebook) | LCC E184.H95 H56 2018 (print) |
DDC 973/.04945110092 [B] —dc23
LC record available at https://lccn.loc.gov/2018007557

CONTENTS

Foreword. Marci Bácsi
CHARLES FENYVESI
ix

Preface. The History of
The Making of an American
CATHY CASSADY CORBIN
xv

Introduction
DOUG CANTRELL
xix

Editorial Note
xxix

---

1. Mátraverebély 1

2. America 13

3. Investment in America 22

4. I Became an American 32

5. Cleveland, Ohio 47

6. Again in Coal Mine 57

7. Riding the Rods 71

8. Romance 77

9. In the Show Business 92

10. I Become a Newspaperman 99

11. On the Road Again 107

12. With Mother 118

13. I Become a Publisher 143

14. I Become a Coal Operator 159

15. Himlerville, Kentucky 169

16. Life in Himlerville 183

17. The Lost Dream 194

18. The American Miner 209

19. Playing God 232

20. In Hungary 239

21. Return Home 245

Appendix 257

Notes 261

Index 295

ILLUSTRATIONS

following page 128

Map of the Proposed Spur at Kermit Station

Map of the Right of Way, Himler Spur,
Himler Coal Company

S.S. Carpathia

Himlerville Construction

Side View of Martin Himler's House under Construction

Himler Coal Co., Inc., Company Store under Construction

Himlerville Residential Houses

Himler Coal Co. Mine Entrance

Inside the Himler Coal Co. Coal Mine

Inside the Himler Coal Co. Coal Tipple

Himler House Dinner Menu Cards

Col. Himler in OSS Uniform

Himler at his Newspaper Publishing Desk, April 1940

Camp Marcus W. Orr, Salzburg, Austria

László Szillágyi, Martin Himler,
Sgt. Bela Herezeg, and Gabor Vajna

Vilmos Hellebronth, Ferenc Omelka, Ferenc Kiss, Peter Hain,
Ferenc Szálasi, Ekno Gombos, Geza Perenyi, Ede Voigt,
Lajos Simsai, László Endre, and Ferenc Kassai

Cover Sheet to Declassified Photos and Martin Himler
Interrogations of Hungarian Nazi War Criminals

Letter Written from Martin Himler to Former Himlerville Residents

Himler's U.S. passport

Himler Later in Life Relaxing at His Los Angeles Home

Hollywood Forever's Cathedral Mausoleum
Where Himler Is Interred

FOREWORD

Marci Bácsi

I WROTE ABOUT Himler's interrogation of Szálasi for *The Washington Post*'s op-ed page published on December 16, 2003. That article is also included here in the Appendix. With both Himler and Borshy-Kerekes long dead, I thought that I did not have to keep my promise of silence forever. My article drew a parallel between alleged rough treatment of Iraqi dictator Saddam Hussein by the U.S. soldiers who found his hiding place and the slap Himler delivered to Szálasi. The editor gave the article the title "Two Men, One Evil."

Colonel Martin Himler lived many lives: coal miner, labor union organizer, peddler, journalist in two languages, founder of a small town that offered housing with middleclass comfort to miners, World War II intelligence officer, and interrogator of war criminals. An unskilled immigrant arriving in the United States in 1908 at age 18 without knowing English, he was an autodidact without a high school diploma. He never had much in the way of formal education, but he was an avid reader of books and a good listener.

His autobiography offers fascinating testimony about the hard times of a little-known immigrant community in the first part of the twentieth century.

In and out of the coal mines of Kentucky, Tennessee, West Virginia, Virginia, and Pennsylvania, Himler ignored the dangers of getting killed by teams of goons employed by the mining companies that claimed to represent the law in rural locations and beat up and even shot people they hunted down as "troublemakers." As a journalist, Himler fought for the rights of miners and other workers abused by powerful bosses who got away with paying low wages or denying payment, using one pretext or another.

Having had enough of working as a miner, Himler founded a successful mail-order business without knowing the first thing about business, bank accounts, discounts, and tax laws. He was often unemployed and much of the time he had no money to buy himself a decent meal. But he helped others less enterprising to get ahead in life.

He responded to the many crises he encountered by growing stronger and wiser.

He was short in stature and not built for the kind of heavy physical work that mining and some of his other jobs required. But he worked hard and long hours.

His autobiography bristles with names, causes, and numbers. He reports each tiny change in the pennies he received for an hour's labor. Unlike other successful newcomers, he does not exaggerate the importance of his achievements. He is a reliable witness.

But when it came to his work in U.S. intelligence, he complied with the rules and he did not name his contacts and the projects he was engaged in. His autobiography does not explain how the fledgling Office of Strategic Services (OSS)—the precursor of the Central Intelligence Agency (CIA)—hired him for the sensitive, complicated job of heading the team that watched his native Hungary and assessed the odds for removing it from its titular alliance with Nazi Germany.

John E. Taylor, the late legendary senior archivist at the U.S. National Archives, had high praise for Himler's reports and especially for his interrogations of 71 pro-Nazi officials responsible for massacring thousands of Hungarian Jews and helping the deportation of hundreds of thousands of others to German death camps. Born into a poor Orthodox Jewish family in the village of Pásztó, Himler was a child when he stopped observing Jewish laws and customs. Later on in the United States, he even converted to Catholicism, a secret he kept from his mother. But he cared for Jews persecuted on account of their faith in pre-war and post-war Hungary. He shed tears for the victims he knew.

I met Himler twice in my life. I was about ten years old when I was helping out in the kitchen and serving food at a party in 1946 in Budapest hosted by my uncle Levente Thury, a member of the anti-Nazi resistance, and his wife Mara, then the editorial director of the newspaper *Demokrácia*, to have him meet others in the Smallholders' Party, an anti-Nazi and anti-Communist group. Unfortunately, Himler's autobiography does not go into the details of his work in the 1930s and 40s to encourage anti-Nazi Hungarians in the United States and in Hungary.

But Jamie Jameson, a senior CIA officer after World War II, remembered Himler as "a smart and hardworking activist, the kind of organizer the Allies needed in countries such as Hungary." When in the 1980s I mentioned to him Himler's name, Jameson did not have to check the files to say how "helpful" and "talented" Colonel Himler was.

Both Taylor and Jameson expressed admiration for Himler's skill and energy in rounding up in the post-war chaos of Germany the leaders of Hungary's pro-Nazi Arrow Cross who were asked by Germany to stage a putsch in Budapest in October 1944. In December, the Arrow Cross leadership fled from Budapest to the Reich as the Wehrmacht retreated.

It took what both Americans called "record time" of two months of Himler's interrogations to collect the evidence later used by Hungarian courts to charge the Hungarian Nazis of war crimes. (Some of them were executed but not all of them were punished as the Communists used a few Arrow Cross fanatics to consolidate their takeover of what soon became a "people's republic" under Stalin's dictatorship.)

"Colonel Himler's interrogations established the fundamental historical facts about the characters of the Arrow Cross leaders," Taylor told me and he did not hide his anger that the original texts translated into English were stolen from the U.S. National Archives. "Yes, stolen," the usually mild-mannered, professorial archivist roared while the official government line was that the documents were only "misfiled" and will be located sooner or later. "The investigation produced nothing," he added, "and we will never know who was responsible for the theft—a Federal crime, a felony."

So far, Taylor has proved right.

The second time I met Himler was in 1957 when I was a new refugee in the United States and a freshman at Harvard. He asked me to call him Marci *bácsi*. (*Bácsi* means "uncle" in Hungarian, which is how young people are expected to address their elders. "Marci" is a shortened form for the Hungarian Márton and the American Martin.) The meeting took place in the Washington, D.C. home of one of his closest friends and supporters, György Borshy-Kerekes, a Protestant pastor who headed for many years the Hungarian Reformed Federation of America and was a respected leader in the Hungarian-American community. He was also a friend of my family in Hungary.

Borshy-Kerekes introduced me to his house guest Himler who took a long look at me before he blurted out: "You don't resemble your Aunt Elza."

I had heard that Marci *bácsi* was one of the several men who admired my mother's older sister Elza, a stunning redhead and an aficionado of poetry. He knew that her life ended in a gas chamber in Auschwitz. So, he asked me about my mother and her siblings who survived and whom he had met at Uncle Levente's party in Budapest.

I also knew that Aunt Elza did not pay much attention to the Hungarian-American who visited our family in the 1930s after he made friends with his colleague, Elza's brother-in-law Levente Thury who probably thought that he should not pass a chance to bring together his sister-in-law, a divorcee, with an eligible bachelor looking for a wife. But my mother was better informed. She had no doubt that her sister Elza would look down on an uneducated man.

I was old enough at age 19 to say nothing about any of this gossip.

It did not take long for Marci *bácsi* to ask me if I knew that he had interrogated Ferenc Szálasi and his underlings in Austria in the summer of 1945.

I said yes, and I knew that I passed a test. Marci *bácsi* took a long breath of a man ready to unburden himself of a secret. He said: "There is something I need to tell you I have not told anyone else." He looked at our host and me. He asked: "Could you keep it to yourselves?"

The pastor Borshy-Kerekes and I said yes.

First, Marci *bácsi* explained that he carefully arranged the setting for his interrogations. An American soldier escorted each prisoner to his office, saluted him, and asked permission to leave the room. He said yes, and told the soldier to stand outside the door and to let no one in.

Colonel Himler was sitting at his desk. Without looking at the prisoner, he said that the interrogation would be in Hungarian and the prisoner must stand and answer all his questions.

He did not disclose what several prisoners whom he had interrogated were saying: They all claimed innocence because all the decisions about the deportation of the Jews and other cooperative ventures with the Germans were determined by Szálasi, "the nation's Führer."

Himler told us that he was stern and impersonal, suppressing all emotion. His job was to gather information and to report back to OSS high command. He said that he barely looked at the people he interrogated.

He began by asking the basic questions: name, place and year of birth. Szálasi answered, then said that as a head of state he was not required to answer any additional questions.

Ignoring that statement, Himler went on to his next question: "What is your profession?"

Szálasi said: "The leader of the nation."

"Suddenly," Himler told us, "I lost my self-control. I got up from my chair, walked over to Szálasi who stood ramrod straight, and I slapped his face so hard that he staggered and almost fell. Without a word, I walked back to my chair and sat down. After he recovered his balance, Szálasi said he was ready to answer all my questions. And he did."

Marci *bácsi* looked at us and said: "To this day I am ashamed of what I did. As an American officer, I should not have resorted to violence with a prisoner of war." He stared at the carpet in front of him. "Do not tell anyone what I have done." He looked at me and said: "But being a friend of your family, I wanted to share my secret with you."

"Marci *bácsi*," I burst out. "You slapped that murderer for my family too, for those who perished like my Aunt Elza and your siblings and those of us who survived. You have done nothing to be ashamed of."

Marci *bácsi* got up, held out his hands. I got up too. We shook hands, and he hugged me and kissed me on both cheeks. "Thank you," he said. There were tears in his eyes.

A pause followed. Then the pastor said softly: "Elza was a beautiful woman. Beautiful."

CHARLES FENYVESI
October 23, 2015

PREFACE

The History of *The Making of an American*

MARTIN HIMLER COMPOSED his autobiography between the years of June 1946–July 1961. These fifteen years mark the time between Colonel Himler's 1946 return to America following World War II and his death on July 8, 1961. Although Himler probably kept notes, perhaps even a journal to chronicle his life, the autobiography most likely was composed at Himler's 1505 West 8th Street residence in Los Angeles, California. He relocated to Los Angeles after returning to America in 1946 and worked as a freelance journalist from his Los Angeles home.

Himler included neither notes nor annotations with the autobiography. The main title of the autobiography was chosen by Himler, but the original subtitle, *The Autobiography of a Hunky*, has been replaced by the editors with the current, more descriptive subtitle. Although the autobiography never was published by Himler, I believe that Himler intended the autobiography for publication if simply for the reason that twice he refers to the manuscript as "this book" in the narrative.

Martin Himler's death was the result of pancreatic and liver cancer. When he passed at Cedars of Lebanon Hospital, Los Angeles, on July 8, 1961, Himler's nephew, Andrew Fay Fisher, inherited Himler's possessions, including the autobiography. Mr. Fisher came to America in 1921 to help his Uncle Martin with work at Himlerville and later became the owner of Himler's most famous newspaper, *Magyar Bányászlap* (*Hungarian Miners' Journal*).

Mr. Fisher also resided in Los Angeles, and the autobiography lived with him there until failing health forced Mr. Fisher to move to Cleveland, Ohio, to be near his nephews and Himler's great nephews, Alex Hertz and Gary Hamos. When Mr. Fisher moved into a health care facility, the autobiography found a new home with Mr. Hertz.

Mr. Hertz and his wife, Clara, both Holocaust survivors, and their son, Sanford, traveled to Martin County, Kentucky, in early fall 2003 to visit Himlerville. Evelynn Cassady, president of the Martin County Historical and Genealogical Society, and society members Jay and Geraldine Fitch, met the Hertz family and showed them Himlerville, the openings of the Himler Coal Company mines, and the railroad bridge built across the Tug River by the Himler Coal Company. Mrs. Cassady and Mr. and Mrs. Fitch treated the Hertz family to lunch at Miss Ida's restaurant in Inez, Kentucky, and a conversation about Himler's autobiography arose. As Mrs. Cassady recalled from the luncheon conversation: "Mr. Hertz was pleased with us and trusted me enough to mail me the autobiography manuscript after the Hertz family returned home." She added that, "Mr. Hertz gave the manuscript to the Martin County Historical and Genealogical Society in hopes of getting the real story of the real Mr. Martin Himler told. I promised Mr. Hertz that I would do all I could to see that this was done."

Mrs. Cassady searched for several years to find an editor who would work on the manuscript on a volunteer basis, assess its publication potential, and provide the needed editorial work. I spoke with Mrs. Cassady via phone in the fall of 2010, and she mentioned to me that "the family of a Hungarian coal miner named Martin Himler, who lived in Martin County in the 1920s, gave the historical society Himler's autobiography several years ago." She further explained that, "I don't know what to do with the autobiography, but it's a really good story and tells all about Mr. Himler's life, Martin County, and the time that Mr. Himler served in World War II." I told Mrs. Cassady that I would love to read the autobiography manuscript and gladly would do any needed editing. My husband and I traveled to Martin County in the winter of 2010 to get the autobiography and bring it to our home in Morehead, Kentucky.

Martin Himler's excellent writing skills were immediately apparent as I began to read the autobiography in 2011. He wrote a stunning and heartfelt message of his life as an immigrant, an entrepreneur, a journalist, and a servant to America and Hungary. Himler provided an important snap shot in time, and I felt that his story was worthy of scholarly publication. Following several years of editing, research, and many discussions with the Himler family, Dr. Doug Cantrell, and journalist Charles Fenyvesi, Himler's autobiography was presented on February 1, 2016, to Thomas G. Wells, acquisitions editor with the University of Tennessee Press.

Mrs. Cassady kept her promise to Alex Hertz. Martin Himler's wish for publication of his autobiography has been fulfilled, and now readers throughout the world can learn about Himler's powerful legacy as they read "this book."

CATHY CASSADY CORBIN
Morehead, Kentucky

INTRODUCTION

DOUG CANTRELL

MARTIN HIMLER'S LIFE, as depicted in his autobiography, was an exciting and extraordinary adventure. Himler overcame poverty and prejudice in his native country, Hungary, migrated to America where he embodied the American Dream as an entrepreneur—a peddler, publisher, and coal baron—and went on to work as an agent of the American Office of Strategic Services (OSS) in economically and ideologically ravaged post-World War II Europe. Himler's beginnings, in many respects, are the same as those of many other immigrants who came to the United States seeking economic opportunity as part of a transnational migration of labor necessitated by the world-wide spread of capitalism and the industrial revolution.[1] This transnational migration of people changed the entire American national economy and had localized impacts on the economies of eastern Kentucky, West Virginia, East Tennessee, northern Alabama, southeastern Ohio, southwestern Virginia, and other areas within southern Appalachia. Although the Appalachian south did not receive the number of immigrants that the northeastern and midwestern states attracted, significant numbers of immigrants migrated to the mountain coal fields in search of employment. West Virginia, according to census records, attracted more than 60,000 immigrants, while mining counties in Virginia, Kentucky, Tennessee, and Alabama attracted a lesser number of foreign born workers.[2] The transnational migration of people from Eastern Europe to build railroads, log the virgin Appalachian forests, and mine the region's coal was a clear indicator that the rural Appalachian economy was not isolated from a developing national and global economy.[3] Eventually, national and multinational corporations would own practically all of the mineral, timber, and other natural resources found in southern Appalachia.[4]

Himler's journey to America was part of a phenomena historians term the "new immigration."⁵ Roughly between 1880 and 1915, approximately twenty million people, mostly from southern and eastern European countries and the Middle East, left their homelands for a new beginning in the United States. Approximately two million Hungarians were part of this migration, and many of them were peasants and unskilled workers who migrated to the United States prior to the outbreak of the First World War. Magyars comprised about 650,000 of the immigrants, and the remaining 1.35 million came from a myriad of other ethnic groups, including Russians, Slovaks, Romanians, Croats, Serbs, and Germans. Political issues that arose after Austria and Hungary united under Emperor Franz Josef I in 1867 largely prompted the Hungarian immigration.

The two countries of Hungary and Austria combined had a population of just over fifty million and a variety of ethnic groups who spoke about fifteen different languages. This mixture did not coexist well, and conflict often arose due to cultural differences between the groups. Religious persecution also posed a problem for inhabitants of Austro-Hungarian lands, especially for Jewish, Muslim, Protestant, and Orthodox Christian inhabitants, because the empire's official religion was Roman Catholicism. The united Hungarian and Austrian government's policy was to suppress non-Catholic sects, and this suppression prompted nearly five percent of the Austro-Hungarian Empire's population to immigrate to the United States. Economic problems likewise prompted many Hungarians to leave their native land. Hungary, like many nations during the last half of the nineteenth century, was experiencing its version of the Industrial Revolution. Hungarian industrialization displaced many rural peasants who were then forced to migrate to large towns and cities in search of employment. Since few jobs were available for unskilled laborers from the more rural areas of Hungary, many Hungarians joined the transnational migration in search of employment opportunities afforded by the United States' more robust industrial development. Franz Josef's government, aware of the many political, economic, and religious problems that existed in the Austro-Hungarian Empire, decided to allow the empire's citizens to migrate in hopes of preventing further strife. As a result of this lenient emigration policy, thousands of Jews, including Martin Himler, Czechs from Bohemia, Slovaks from northeastern Hungary, and Poles migrated to the United States to flee religious persecution, avoid the policy of Magyarization imposed by the government, evade military conscription, or escape a dying agricultural subsistence economy.[6]

During the last decades of the nineteenth century, Eastern European migration represented a seismic shift in the sources of immigration to the United States. Prior to 1880, most immigrants on American shores journeyed from northern and western European countries. Not only did the new Eastern European immigrants represent a change in immigrant origin and ethnicity, but they also marked a shift in religious practices. Many of these immigrants were of the Roman Catholic, Jewish, Eastern Orthodox, or Muslim faiths when they entered the United States, which prior to 1880, had been largely a Protestant nation. This shift in ethnicity and religion reignited a nativist movement within the United States that had confronted earlier generations of immigrants.[7]

In the 1920's, nativism resulted in the passage of the National Origins Act (NOA), a restrictive legislation designed to limit the numbers of southern and eastern European immigrants allowed in to the United States under a quota system. Even more restrictive measures were enacted when the NOA was revised in 1924 and again in 1927. Under this restrictive quota system, only about eight hundred Hungarians were permitted entrance into the United States each year. Himler starkly confronts this nativism within the pages of his autobiography. Although Himler's motives were complex, the establishment of cooperative coal mining communities, first in West Virginia and then in eastern Kentucky, was in part a reaction to the prejudice, discrimination, and racism that he and other Hungarians faced. Certainly an ancillary goal of Himler's establishment of mining communities owned and operated collectively by Hungarian immigrants was to demonstrate to American nativists that immigrants were productive people who could work and prosper within the larger American capitalistic system. By harnessing capitalism, industrialism, technology, and labor, Himler hoped to convince Americans concerned by the influx of foreign-born people that immigrants could contribute to the American laboring class and that immigrants did not pose a threat to American capitalism, industrialism, religion, or democracy. Following the bankruptcy of the Himler Coal Company, Himler's subsequent work with the Chicago, Milwaukee, St. Paul, and Pacific Railroad to establish Hungarian agricultural colonies in the Midwest was dually an attempt to thwart nativism and to demonstrate that immigrants could assimilate into American agricultural communities.

Though Himler converted to Catholicism in America, he left his native Hungary as a member of the Jewish faith. Because of the caste system present during the late nineteenth century and the mistreatment of Jews in Hungarian society, Himler, like many other Jewish immigrants, sought refuge in American religious tolerance. Of course, Himler also encountered prejudice and discrimination in the United States. He did not naively believe that American streets were paved with gold; he understood that in order to be successful in America, he would have to work for that success.

During Himler's early years in America, he found employment at a number of arduous manual labor jobs. These jobs included such diverse tasks as mining coal in West Virginia and Pennsylvania, carrying molten iron in a steel mill in Ohio, building the McAdoo tunnel in New York, repairing shoes in a cobbler's shop, building sand molds to make iron products, washing windows in a hotel, and serving passengers on a ship. He even trespassed from immigrant community to immigrant community throughout the Appalachian South as a pack peddler carrying suitcases filled with merchandise needed by foreigners working in mountain coal camps.[8] Himler initially did not come to the United States with the idea of permanent residency. His idea at first was to come to America for a temporary stay until he earned money enough to return home and live in the lifestyle that people of his caste were required to occupy. Immigrants of a similar mindset as Himler usually had to abandon their plans of returning home after facing the reality of low wages and harsh working conditions that left them little choice but to remain in the United States.

Historians throughout the years generally have viewed immigration from several different perspectives. Early scholars described immigration as the desperate flight of impoverished people who left their homelands in search of the American Dream. Once immigrants arrived in the United States, however, this view held that they were victimized by the capitalistic system that forced them to labor hard in low-paying jobs and live in urban ghettos. Later, scholars depicted immigrants to the United States as cultural traditionalists who created unique institutions, such as the ethnic community in the urban ghetto, foreign language newspapers, and mutual beneficial societies that allowed the newcomers to rely upon their old world traditions to adjust to life within the American economic, industrial, and social system. Yet another scholarly interpretation holds that capitalism uprooted immigrants from the traditional society and culture in their native lands and imbued them with a desire to seek the benefits of capitalism in the United States, which, by 1880, was perhaps the most capitalistic of all countries. Although these interpretations differ about why immigrants migrated to the United States, they generally hold one assumption in common—all immigrant groups shared largely a similar experience in America with only slight differences.⁹

John Bodnar challenges this common assumption of earlier immigration historians in *The Transplanted: A History of Immigrants in Urban America*. He maintains that the immigrant experience cannot be so easily categorized. Not all immigrants left their homelands for the same reason and not all had the same experiences once they arrived in the United States. Each immigrant, Bodnar and recent scholarship suggest, was an individual who left their homeland for reasons unique to the person and who confronted life in America on his or her own terms. Some immigrants held on to traditional life while others quickly assimilated into American life. Martin Himler's story seems to prove Bodnar correct. Himler, as he denotes within the pages of his autobiography, embraced capitalism and unleashed his entrepreneurial spirit. He opened a business as a pack peddler, established *Magyar Bányászlap* and other publications, built two cooperative coal mining communities, and rubbed elbows with several leading capitalists in Kentucky, West Virginia, Ohio, and throughout the nation. Himler's rejection of radical ideologies on both the left and right of the political spectrum caused ideological extremists within the immigrant community to attack him. Hungarian immigrant leaders who had adopted more radical philosophies often criticized Himler and his entrepreneurial ventures.¹⁰ For example, the radicals ridiculed Himler's creation of *Magyar Bányászlap*. They called Himler a "parlor miner" and sarcastically asked why Hungarian miners needed their own newspaper. Writers of the radical publications often maintained that Himler was one of many crooks and scam artists who fleeced immigrant communities. When the Himler Coal Company, the corporation that controlled the cooperative mining enterprise at Himlerville, went bankrupt due to depressed conditions in the coal market and a progressive thickening of a slate layer within the Himlerville coal seam, radical immigrant leaders in the United States crowed that Himler had cheated thousands of hard-working Hungarians out of their life savings. Himler refutes these allegations in his autobiography by explaining that when he left Himlerville following the bankruptcy to be treated for oral cancer at the Mayo Clinic in Rochester, Minnesota, he had only four dollars in his pocket and had to borrow funds from friends to cover living expenses. Even today, rumors that Himler absconded with thousands of dollars stolen from immigrants who invested money in the Himler Coal Company persist in Martin County, Kentucky, and among some descendants of former Himler Coal Company investors.

Himler's rejection of radical political and economic ideologies and the resulting attacks upon him, along with nativism, prejudice, and discrimination, help explain Himler's fight against Nazism and Communism after World War II and during the Cold War period. Himler became an ardent foe of Communism, and, at least in his view, his greatest success occurred after he enlisted in the American military following the Japanese attack on Pearl Harbor. Himler joined the army at the relatively advanced age of fifty-five because he wanted to protect the country that had enabled him to succeed and because he believed that Nazi ideology was a threat to the American, Hungarian, and European way of life. He and the thousands of other immigrants who enlisted and fought for the United States against the German and Japanese militaries wanted to challenge ideas that immigrants were not patriotic.

Since Himler was fluent in several languages, including Hungarian and English, he was assigned to the Office of Strategic Services (OSS), which later became the modern Central Intelligence Agency (CIA). As an OSS officer, Himler was sent to various locations in Europe and Africa where he helped civilian refugees survive in their war ravaged nations, distributed food and clothing, and worked to provide shelter to the homeless. Part of Himler's duty with the OSS involved investigating, arresting, and interrogating Nazi war criminals. Unfortunately, several members of Himler's Hungarian family were among the 600,000 Hungarian Jews who perished in Nazi extermination camps. Although Himler's words in the autobiography deny any personal motive in sending these Nazi brutes to face trial and execution, he understood that justice was being done, especially after learning that some of these criminals likely had a hand in the extermination of members of his own family.

Himler not only rejected the right wing extremism of Nazism, but following the end of the Second World War when the Soviet Union posed a threat to the United States, he also rejected the left wing extremism of Communism. He lived through the Cold War as an ardent foe of Communism, writing his autobiography at its height and as an agent of the OSS. Throughout the pages of the autobiography, Himler condemns Communism as a grave threat to American democracy and to the capitalistic system that, in Himler's view, had created within the United States a land of vast economic opportunity. Himler's objection to Communism was, in part, a reflection of the time in which he wrote. Himler likely feared that American ideologues such as Senator Joseph McCarthy, Richard Nixon, and others, in the context of the Second Red Scare that erupted at the beginning of the Cold War, would tar immigrants with Communism and produce a second period of nativism similar in scope to the 1920s following World War I.[11] Himler did not want to battle racism, prejudice, and discrimination as he had earlier in the coalfields of West Virginia and eastern Kentucky.[12] Himler also was aware of the precarious situation in Hungary following the Second World War when the Soviet Union dominated Himler's homeland. Approximately 130,000 Hungarians migrated to the United States to escape Soviet rule. Nearly 30,000 of these came as displaced persons, and about 40,000 were Hungarian freedom fighters who escaped after the Soviet Union suppressed the Hungarian Revolution in 1956.

Martin Himler's ideas on American capitalism and industrialism were shaped to some degree by the Progressive Movement and the Great Depression.[13] Himler, who arrived in the United States in 1907, came when the Progressive Movement was at its height under President Theodore Roosevelt who, along with his successors William Howard Taft and Woodrow Wilson, worked to strengthen American capitalism through the regulation of industry. While some of the Progressive regulation was designed to right previous wrongs and injustices, much of it was done at the behest of industrialists themselves who wanted the government to level the playing field for all. Like most Americans, Himler suffered during the Great Depression and generally supported Franklin Roosevelt's New Deal policies and regulations. Himler barely managed to keep *Magyar Bányászlap* financially solvent due to the loss of advertising and subscription revenue as the Great Depression closed businesses and factories, causing massive unemployment that in turn caused thousands of Hungarian readers to cancel subscriptions after they had lost their jobs. Although Himler recognized that many of Roosevelt's ideas were socialistic, he understood that the genius of the New Deal was that it used socialism to preserve the capitalistic system within the United States.

Following Himler's discharge from the military in 1947, he moved to California where he hoped to live the life of a simple farmer, writer, and publisher. Unfortunately, Himler's business partner and friend, Johnny Bako, died before their dream of living as gentlemen farmers was realized. Himler spent the remainder of his days in Los Angeles where he worked as a freelance writer until his death in 1961 from pancreatic and liver cancer.

Himler's autobiography reads like a life well lived. From his earliest travels in Hungary to his precarious passage across the Atlantic, and from his joyful landing in America to his rise as a southern Appalachian coal baron and newspaper magnate, Himler openly discusses his hopes, fears, dreams, and trepidations. His career as an unlikely OSS officer is no less than utterly fascinating, and his skill with multiple languages and interrogation techniques helped convict war criminals who were responsible for one of the greatest genocides in human history. Martin Himler may not embody the American Dream in all its multifaceted definitions, but his literary legacy can teach Americans much about immigration, the melting pot, the people and economics of Appalachia, and the horrors of the Holocaust and the Cold War era of capitalism versus Communism.

Readers of the autobiography will enjoy a story that reads like a novel, that holds their interest throughout, and that provides a useful and unique view on immigration from the perspective of an immigrant. Himler's struggles in surmounting prejudice and racism, overcoming poverty, and his successes and failures in his effort to become American will give hope to all peoples, both immigrant and native born, who are fighting to overcome class and caste, hatred, discrimination, and other obstacles encountered in life. Himler's life vividly illustrates that if people are unafraid of failure and are willing to take risks and continue despite life's setbacks, obstacles can be overcome and success awaits those who endure and who continue the struggle.

Himler's life certainly was a success. He escaped the class and religious prejudice in Europe, encountered a different kind of prejudice in the United States, and rose from the ranks of an ordinary laborer to achieve success as a businessman and publisher. Himler could have been content to rest on his laurels and live out his life in relative comfort, but he felt his duty, despite his advanced age, was to defend his adopted nation against its foreign enemies. He certainly achieved the American Dream, embraced American ideas, and in the end, became fully American. Yet, Himler never forgot his immigrant past, worked continually throughout his life to improve the lot of his countrymen and all immigrants, and served his adopted country proudly during the Second World War and in the early days of the Cold War. As an American, Martin Himler will be remembered as one of the United States' great immigrant success stories, and his story will offer hope to future generations struggling to achieve the American Dream.

EDITORIAL NOTE

MARTIN HIMLER WAS a skilled writer and a natural storyteller. His animated personality, wit, and keen knowledge of his subject matter engages readers in a literary conversation just as these qualities no doubt engaged face-to-face conversations with individuals to whom Himler spoke.

The editing of Himler's autobiography made no changes in what Himler said or in the information that Himler wanted his readers to know. The only changes made involved usual editorial work, such as structure and grammar. Himler had a tendency to write run-on sentences, so sentence structure was improved to give readers the full impact of Himler's words. Verb tense required work, but Himler's use of tense was excellent when one considers that English was Himler's second language.

The autobiography also needed some work to ensure that paragraphs weren't too long or too short and that paragraph subjects were well divided. Edits were made to correct unclear antecedents, give parallelism when needed, and to correct spelling and punctuation. Himler used several exclamation marks in his autobiography, and it's very interesting to read the topics and events that he felt were worthy of the exclamation mark.

Martin Himler was a tireless advocate for the underdog, and the autobiography often finds him explaining the hardships and disadvantages facing minority groups and immigrants in America. As he discusses minority groups, Himler uses terms such as Hunkies, Negroes, Hunyaks, Dagoes, and hillbillies, all of which had a wide currency in his lifetime. Martin Himler's use of these terms was not derogatory. On the contrary, conscientious readers will quickly see that the words come from a heart full of kindness and love for people of all races and ethnicities.

Colonel Himler's autobiography doesn't provide extensive detail about his Office of Strategic Services work. His work was classified information at the time the autobiography was written and wasn't declassified until many years later. Himler's documents and papers that are filed at the National Archives, College Park, Maryland, are stamped SECRET in red ink.

As I edited the autobiography, I sometimes felt that Himler wrote in a bit of a hurry. Perhaps he was trying to get all of the information written as soon as possible so that he could seek a publisher and bring the autobiography to publication during his lifetime. Although Himler's autobiography wasn't published till almost sixty years after it was written, Himler's message remains an important, relevant, and stunning account of American, Hungarian, and world history.

Martin Himler's Dutch Colonial Revival style home, Himler House, stands on top of a mountain overlooking Beauty, formerly Himlerville, Kentucky. Himler House received a listing on the National Register of Historic Places in August 1991, but at this time, the picture of Himler's life was still fragmentary. The Himler autobiography brought new information and research opportunities and helped to clarify existing information about Himler Coal Company, Himlerville, Himler's journalism career, and his service to America, Hungary, and the world as an OSS officer during World War II. A restoration of Himler House, coupled with the publication of Himler's life story, will give Himler House the potential to earn the designation of a U.S. National Historic Landmark dedicated to Appalachian coal mining history and Hungarian immigrant culture. Therefore, all royalties from the publication of the Himler autobiography will be donated to the Himler House Restoration Fund, Martin County Historical and Genealogical Society, Inc.

THE
MAKING
OF AN
AMERICAN

★

CHAPTER 1

Mátraverebély

★

There were lots of people in our home in Pásztó[1] on a fall day in 1892. My oldest sister, Terez, was dressed up in a beautiful pink silk dress, and she was sitting in an upholstered chair borrowed from Mrs. Gzobel, the wife of the rich Jewish merchant. Terez married Adolph Fischer, an insurance agent.

There were large plates in the kitchen loaded with roast geese, duck, and strudel. I snatched a piece now and then. Finally, I had to go out to the yard and throw up. I felt terrible, went to the stable, and fell asleep. It was very late in the evening when my family found me, and I was told that the wedding was over. I missed the show.

This was the first day of my life that I remember. I was four years old. Next morning, our furniture, pots, and pans were loaded on two wagons. My parents climbed on the first, my two sisters, my youngest brother, and I on the second, and then we were on our way to our new home in a village eight kilometers away, a village called Mátraverebély.[2] My brother and I rode on top of a big box sitting on the wagon.

There were thirteen of us in our family which included my father, an old and by that time semi-invalid man, my mother, an illiterate but wonderfully wise and intelligent woman, and eleven children. The oldest of the eleven was Terez, the bride, and the next in line was Herman, locksmith, then Morris, a dyer, Terka, a dressmaker, David, a tinsmith, Charlotte, home with my parents, Jakab, a cabinet maker, Bertha, a dressmaker, Regina, home with my parents, Géza, a schoolboy, and I, the youngest child. Herman, Morris, and David were journeymen in Budapest and supported themselves, and Jakab was an apprentice, also in Budapest, and Terka and Bertha remained in Pásztó to board with the newlyweds and support themselves. Charlotte, Regina, Géza, and I went with our parents to Mátraverebély.

My family was to operate a korcsma in our new home. Our home was owned by the other Jewish family in the village who, having forty acres of land and a grocery store, was too rich to bother with the korcsma, the equivalent of the American saloon.

Géza, two years my senior, and I were fascinated with our new home. It consisted of a very large hall with a small place partitioned off for the drinks, a large kitchen which also was the dining and living room, and the clean room. The clean room held all the good furniture and all the little treasures that make up a parlor and also served as the bedroom for my parents. We only went to the clean room during the day if out-of-town guests were entertained.

There was a large basement to keep the beer, whiskey, and wine cool, and on Sundays, we had to run down there innumerable times for a bottle of beer or a pitcher full of whiskey or wine. We also had a large yard, a tremendous garden, and a stable. The stable was used to house a goat for my father was ordered by doctors to drink goat milk.

Exploring the home and its environment and getting acquainted with the kids of the village kept me pleasantly occupied even after Géza was sent back to Pásztó within a few weeks to attend school. Géza also was boarding with Terez.

Nominally, of course, the business was managed by my father, but the real manager was Charlotte who was about nineteen years old. My mother took care of the household, cooking, milking the goat, patching, scrubbing, gardening, and attending to the many needs of my father. Regina helped my mother. But, when Charlotte was busy in the korcsma, especially on Sundays, all of us had to assist her. I did not play outside with the other kids on Sundays; I spent most of my time in the korcsma. On weekdays, during the day, we had very few customers, and the korcsma, with its very long tables and benches, served as a dining room as well as our living room.

I especially enjoyed the evenings. There were a few steady customers who always dropped in, and when they were through discussing the weather and the crops, I was fascinated by their stories about ghosts and witches as described by those who had seen one of the two of them. Whenever someone died in the village, ghosts were subjects of lively discussions.

It was customary to ring a bell every four hours in honor of the dead, and at the sound of the bell, relatives and friends were supposed to visit the house of mourning. The body was laid out in the middle of the room, and close female relatives bent over it. The females cried and enumerated the merits of the dead and the great loss of the family and included an occasional kiss for the dead one. This mourning continued while the callers respectfully stood around the dead one, and the mourning ritual was repeated four times a day until the funeral.

But there were also more joyful occasions for semi-public gatherings. When winter set in, womenfolk got together in the evenings to make thread of flax, and they were joined by the young men to entertain them with stories and songs. They usually had a jolly time. Kids were permitted to attend as long as we behaved.

If a girl were steadily escorted home by one of the male youngsters, it meant serious intentions. In such case, the boy slept at the girl's home in a separate room with the girl, and while supposedly they were only petting and necking, I know of only two cases during about twenty years where the trust was violated.

On Sundays, the social life of the village was centered in the korcsma. By about 1:00 P.M., the youth of the village entered in bunches, occupied the longest table, and by 2:00 P.M., there were about twenty to thirty of them drinking and singing. Older men sat at another table. Very few got drunk for they were too poor to spend much money, except on important holidays such as when a youngster was accepted as a grown-up person.

All the grown-up persons were organized into three groups with such groups contracting to harvest the wheat of Count Almassy who owned nine tenths of the land around the village. Girls and youngsters were accepted to gather and bind the wheat into bundles, and they received half shares of wheat for their work. And when a young man reached about seventeen to eighteen years old, he was accepted as a harvester and received full shares for his work. This meant that he was permitted to consider himself a grown-up young man and to sit in with the rest on Sundays after buying two liters of whiskey for the group.

Following supper on Sunday, the young men were joined by the girls, a few gypsies arrived, and the dance was on until about 10:00 P.M. Occasionally there was a fight, even a little knifing, but not often.

Mátraverebély was a very poor village. The land was owned by Count Almassy, with most of the peasants having two to three acres, enough to raise some potatoes, corn, etc. The three families having over ten acres were considered wealthy. During the summer, they hired out workers to hoe sugar beet and corn for the Count, and they also tended their own patch of land. A few peasants were employed by the railroad as section hands, and the very poor went to work in a stone quarry or a coal mine about eight miles away. The poor walked the distance to the quarry and the coal mine twice a day.

Since most of the able-bodied men were harvesting for the Count, lots of the work fell on the women during July and August. Harvesters worked from 3:00 A.M. until about 9:00 P.M., and they did not go home to sleep. The women had to hoe the potatoes and cabbage, and they had to take care of the cow, the children, the pig, and the chickens. It was especially hard on the younger women because babies usually were coming around August and September as a consequence of the idle days of the winter months.

The Mátraverebély families had very little to do during the winter except the small chore of tending a cow, and the families who were moderately well-off tended to their horses. Among the approximately two hundred Mátraverebély families, only thirty-two families had horse teams. These families did the plowing and hauling for everyone else. On weekdays, the men did odd work around the house in pants turned inside-out. On Sundays, the pants were worn with the regular black color outside.

Most of the families were too poor to afford three meals during the winter months. They got up late and retired early, thus doing on two meals and saving kerosene. Very few lights were burning in the village after 7:30 P.M. during the winter.

Many of the families managed to have one hog and killing and preparing same was a holiday with close friends and relatives invited to assist. The hog had to serve as meat and bacon during the whole year, and if the hog died of some sickness, it was a catastrophe for it meant semi-starvation for the year. Bread was provided by the harvester's share, received in kind, and ground in the water mill at Pásztó.

Since the village was poor, we were poor. Very poor. For years, I did not have a new pair of pants but had to wear those out-grown by Géza, and my pants usually were decorated by several patches by the time they reached me. But, just as the village folks were satisfied with their lot and had a comparatively contented life, poverty did not bother me much in my childhood. I never went hungry while home. Mother baked enough black bread each Friday to last a week. We had coffee and toast for breakfast, a square meal at midday with meat or poultry at least four times a week and for supper, I always received four walnuts or a little home-made jelly with the bread. Friday night, we had white "barches" and a hot meal; Saturday, we had cheesecake and usually roast duck or goose.

But with all that, I was unhappy most of the time. My father was building a ghetto around me. He was an Orthodox Jew,[3] and he made me miserable with all the restrictions he imposed on me. I was the only Jewish boy in the village. Most of the four, five, and six year-old boys wore only a shirt, and the older boys wore a "gatya" (skirt-like underwear). However, as a result of my father's orders, I had to go around in pants. Sheer poverty forced my father to let me go barefooted, but he was very unhappy about it. Since my father was sick, he was moody and impatient. He scolded me almost daily for playing with the peasant kids. I never knew any other quarrels between my parents except those occasions when my mother always came to my defense. Even then, my parents quarreled in Yiddish,[4] a language I never mastered.

I did not enjoy being dressed in my best suit and wearing shoes on Saturdays, for I "clicked" with my playmates on those days even less than on other days. I was unhappy every Sunday when my playmates were dressed up. When I induced some of my buddies to play in our yard, mother usually handed out a piece of bread for every one of us, but I was not permitted to accept similar treats elsewhere. Of course, my friends could not help to note the difference, and the minute we quarreled, I was called a "ghinny" by some of them.

When I was six years old, I was sent to the village school for we could not afford to pay board for both Géza and me. It was a Catholic school and, of course, I never was called on at the start to recite the Lord's Prayer or at the close to recite the Hail Mary.[5] I knew them better than most of my fellow students, but the teacher was tactful, and I became more and more an outsider.

The following spring, my older brothers came home for Easter, and they were not satisfied with the progress I had made in the poor village school. My father told them about my becoming a regular "goy" (Gentile)[6] under the influence of the Catholic school. The four journeymen brothers agreed to pay my monthly board, 12 Kronen, $2.40 monthly, so that I could return to Pásztó and attend the excellent Jewish grade school there. In September, I was sent to Pásztó where there was an excellent Jewish grade school. I had to repeat first grade. Perhaps because I had the advantage of some previous schooling, I was an excellent student for the first day and until I was expelled from a Normal School. But my father was worried about my Catholic training and nothing would do but hiring a "bocher" (a student of Torah[7] preparing to be a Rabbi) to teach me Hebrew. It meant an hour wasted every evening, and I deeply resented the time I missed playing. Géza finished the compulsory four grades at the close of the year. My older brothers decided that four skilled workers were enough for a family and that Géza and I would receive higher education. Géza was put to high school[8] in Pásztó.

My life in Pásztó was not very happy. Besides spending an hour with the Hebrew teacher, I had to nurse Terez's children, run errands for Terka and Bertha, my two dressmaker sisters, and the whole atmosphere was not happy. My brother-in-law was a gambler and often lost his earnings. When my sister quarreled with him or when there was hardly enough food on the table, he usually wound up the quarrel by throwing up the fact that he married four to five persons instead of one. As a matter of fact, his family would have starved without the income from the boarders.

During the summer vacations, Géza and I had a fair time, but he was a quiet boy, and I was wild. He spent lots of his time reading at home while I was roaming around with my friends and getting scolded by my father more and more often.

Four years later, I also was enrolled in the high school and was still boarding with Terez. I finished the first year of high school with flying colors when Géza was graduating from high school. My family could not get over the fact that while Géza was a very good boy and did his homework every evening, I remained wild, never did any homework at all, yet finished my first year of high school as the best student in the school. I thus received a 10 Kronen gold piece for my achievements, but Géza received only 5 Kronen.

A bold decision was made by my brothers. Géza was sent to Turócszentmárton[9] to enroll in the Academy of Commerce. My brothers were to furnish money for tuition and the necessary incidentals, and Géza was to "eat days."

This "eating days" was an old institution amongst Jewish people. Poor students went around to the Jewish families of the town and requested the families to furnish three meals one day of the week during the year. Most of the families were willing to accept a student for a day each week, some well-to-do families accepted one for every day of the week, or even several for one certain day. When the student was assured meals for every day, he was set. He went Sunday to X family for meals, Monday to Y family, and so on throughout the year. In some cases, the student was able to receive meals year after year from the same families. Of course, it also happened that some students were not able to get fed more than four to five days a week; other days, they had to take care of themselves. While it was tough to accept charity day after day, three times a day, many outstanding professionals in Hungary were assisted through schools by this "eating days" institution, and Géza was one of them.

When I was about to start my third year in high school, the Royal Hungarian State Railway started to sell student passes at a low price, and a pass from Mátraverebély to Pásztó was sold for 3 Kronen. Meantime, my sister Terka was married to a tailor, my brother Morris also married, and both settled in Pásztó. Since we were all fed up listening to my brother-in-law's complaints as to how he married five or six of us, Bertha went to board with Morris and a commuter's ticket was purchased for me. I was very proud to be photographed for the pass, and Morris's wife lent me a necktie for the occasion.

The train left Mátraverebély at 7:00 A.M., and we just had time to reach school at about 8:00 A.M. But, returning home was more complicated. There was no train until 9:00 P.M., and it was close to 10:00 P.M. when I reached home. School was out at 4:00 P.M., and I was supposed to spend the time between school and the train at Terez's home to do my homework during the winter and to walk home during the summer with another boy of Mátraverebély, the son of the station master.

Sometimes, I did go to Terez's home and help her with the children. More often, I played at the railroad station for there were about fifteen of us commuter boys, and we caused a good many grey hairs to the railroad workers. We were a bad lot, indeed.

The last straw on the ever-strained relationship between my father and myself came about the first Saturday of my commuting days. I was ordered to walk both ways for Jews are not permitted to travel by train on Sabbath.

Thus, I became once more an outsider, even amongst my closest friends who were kidding me plenty. They did that especially on winter days. My mother could not see my walking in six inches of snow or heavy rain in the dark mornings, and she instructed me to leave the house at half past 5:00 A.M. and start toward Pásztó, but then turn back outside the village and go to the station through the pasture. Since school was out at noon on Saturdays, Mother advised me to take the 1:00 P.M. train as far as Tar, the village between Pásztó and Mátraverebély, and walk home from there, thus saving half the distance. My schoolmates knew very well my reasons for getting off at Tar, and they kidded me mercilessly.

And I knew then that I never would have a Jewish child; I never would force the terrible inheritance of being a Jew on my children. In my rebellious mood, I reasoned out that it is more of a work to walk than to ride trains, and I found lots of inconsistencies in my father's belief.

When we had to call on some kid to put wood on the fire on Sabbath days during winter, I reminded mother of the Jewish Ten Commandments which forbade Jews to work or to have their servants or servant girls work on Sabbath days. Mother did not argue with me, and I would not think of arguing with her or even mentioning such matters to my father. In fact, I never spoke to my father unless I had to.

During my 1901 high school year, I got typhoid fever and was laid out in Morris' home for six weeks, most of the time unconscious. One of the doctors pronounced me dead, and as is customary with Orthodox Jews, I was put on the floor and covered by a black sheet kept for that purpose by the Chavre Kadishe, the burial organization of the Jewish community. Word was sent to my school, and students were instructed to appear next day in their Sunday suits for my burial. While they were praying over me, I moved, and I was quickly put back to bed. This was a turning point.

I recovered slowly, but I was so weak that I had to learn again to walk. Since there were some intestinal complications, my family was afraid to feed me. I was kept alive for weeks on wine and cognac,[10] for I would not keep milk. But I was very hungry and grew to hate the very smell of wine and cognac. To this day, I do not like to drink alcohol.

When I was well enough to be carried home to Mátraverebély, I was told the whole story of my illness. I was amazed to hear that once, when I was unconscious, I had asked for my father who rushed in from Mátraverebély crying. Then I understood that I was starved for the love and companionship of a father and, from then on, our relation was more friendly, or at least less strained.

Upon my return home to Mátraverebély, I was much amused by the older peasants who kept on questioning me about the after-world. They were kept posted during my illness, they knew that I was "dead" for about two hours, and they would not believe that I did not meet some of the previously departed Mátraverebély folks in the after-world.

By now, I was thirteen years old, and it was time to be "Bar Mitzvah," which means confirmation in the Jewish faith. I did not wish to be confirmed in a faith I did not possess, and I told my mother so, frankly. She was happy about my decision. She loved my father dearly and accepted his dictum in everything, including the Orthodox faith and life. But, she also was the most understanding mother a son could have.

My mother then convinced my father that our family couldn't afford the expenses of providing a new suit for me and the customary confirmation party in Pásztó, and therefore, the ceremony would, as of necessity, have to be delayed. About eight months later, my father died while Géza and I were in Turócszentmárton, and our family was too poor to have Géza and me attend our father's funeral. The matter of my confirmation was never mentioned thereafter, nor was I ever confirmed in the Jewish faith.

Turócszentmárton was a Slovak town, and Géza learned some Slovak. I also wanted to learn Slovak, and my family agreed that I would enroll for my fourth year of high school in Turócszentmárton. During the summer vacation preceding my going to Turócszentmárton, I went to work for my oldest brother in his small grocery store in Mátraverebély. My brother also had rented a thrashing machine and operating the machine was hard work for me. I was not quite fourteen years old. My mother was opposed to the thrashing work, but money was needed badly, and I was permitted to hire out. I had to work just as the others, from 5:00 A.M. until 8:00 P.M., and I received the same pay, 80 Filler, $.30. But, I always was happy when we were rained out.

In September, I arrived in Turócszentmárton and went around to the Jewish families to try to secure meals. Both Géza and I were unlucky; we could get only five day meals per week. Each of our older brothers contributed so much per month, hardly enough to pay for room and provide incidentals, so we had to stretch our joint income to provide food for the additional two days per week. Our budget allowed 32 Fillers per day for food. Usually, 20 Fillers were spent for a bread, and 12 Fillers for cheese or a herring or, for me, because Géza would not touch it, a small piece of bacon. But, when Géza and I were faced with some contingency—for instance, when shoes had to be heeled or re-soled—our budget became woefully unbalanced, and for weeks, bread had to suffice for our meals on empty days. We did not wish to worry our mother and our family was unaware about our semi-starvation, but I often cried because I was hungry.

We had a room with a Slovak family and at the end of the year, all members of the family spoke some Hungarian, but I failed to learn Slovak. Languages were, and are, one of my weaknesses. Although we had to learn German in the Jewish grade school, as well as in the high school, I never mastered the language. I failed to learn Slovak, and it took me five years to understand English in America.

At the end of the school year, it was decided by the family council that I was not to follow Géza in the Academy of Commerce. My older brothers and sisters were just as aware of my antagonism to everything Jewish as my father was and to offset this strange streak in me, I was to go to the Jewish Normal School of Budapest. This decision was made all the more so because Géza had graduated, and I was too wild to be in far-away Turócszentmárton alone. While in Budapest, I would be under the supervision of Jakab who just about the same time established his own small shop as a cabinetmaker. Of course, I was to "eat days," and meals for six days were provided by distant relatives. The seventh day, I went to the People's Kitchen, a charitable institution where meals were sold for 20 Fillers.

The Jewish Normal School was supported by Jewish philanthropists, and needy students were provided with tickets to the People's Kitchen for midday meal and also with 6 Kronen a month for room. I applied for and received tickets for seven days and also the 6 Kronen for lodging. I also secured "Positions" to tutor stupid and spoiled brats, and I received 2-4 Kronen a month for each of them and about 10 Kronen a month from my brothers. As long as I went to "eat days," I was financially well off for I sold six People's Kitchen tickets each week to the poor old Jews at 16 Fillers each and used the proceeds to attend the theatre. But "eat days" were very humiliating and in a few months, I dropped all but two of them.

By forcing me to the Jewish school, the family did the worst thing. I hated the school and attending same alienated me forever from the Jewish religion. I was not interested in Hebrew, nor in the history of Jewry, and I skipped lessons. Instead of attending such lessons, I went to libraries, the parliament, and court sessions. Jakab did try to supervise me, but Budapest is a large city, and Jakab was busy with his new shop. He seldom knew what I was doing. Truancy became a habit with me, and I only attended hours of my favorite subjects. If I were a wild boy before, I became a bad youngster in Budapest where all the temptations of a large city surrounded a boy in his fifteenth year.

The Director of the Jewish School called me into his office at the end of the year. He told me that I was the worst student, that they would not put up with me any longer, and that I would be expelled. No explanation or begging was accepted. He said that since I seemed to have a brilliant mind, I could have been the best in school, and the stiff lesson of being expelled might wake me up to myself. Or, if I were rotten to the core, the expulsion might just put me on the road toward which I was headed.

I was the best student of Hungarian literature in the whole Jewish school. They had no choice but to award me the prize of 10 Kronen donated expressly for that purpose. But at the same time, it was announced at commencement that Martin Himler failed in five other subjects and was expelled. The Director was right. I was awake and aware of my faults. I was very grateful, and in my later years, the Director and I became very warm friends.

It was not easy to return to my folks, especially to my mother, although she was the only member of the family who did not lecture me. But far more terrible than any lecture, she received me crying. My brothers and myself were for an apprenticeship at some trade, but mother insisted that I become a "learned man" and that I had to return to the fifth grade of high school and once more commute to Pásztó.

Meanwhile, Géza had a good position as clerk in a manufacturing plant in Losonc[11] and a half year later, he secured me a similar job with a stamping and enameling factory. Thus, at the age of sixteen years, I had a good job usually awarded only to a graduate of the Academy of Commerce, and I received a good pay, enough to pay for my board and clothing befitting an office worker. Géza and I were boarding with a family and lived on a scale we never experienced at home, with a hot meal even for supper.

My mother felt that she was justified in her faith in me, and my sisters and brothers were satisfied. A half-year later, they had to witness the same phenomenon they wondered at during school days. Géza, the model young man, was getting a pay of 80 Kronen a month. I, the black sheep, was getting 90 Kronen. Géza was happy about this, for he was somewhat fearful in recommending me to the job, but I felt like two cents. I loved Géza very much, and I was very proud of him.

About a year later, when I was seventeen years old, I fell in love. The owner of the building in which we lived had two daughters, and we were together a lot. Géza hardly knew them for they were Gentiles, but I spent all my free time in their apartment until I discovered that I was desperately in love with Icza.

My love was returned, but we knew that it was hopeless. Icza and her family were rich, and I was poor. They were Gentiles, and I was a Jew. We were much too young to think of early marriage, but we could not see any hope even in the distant future. If I were lucky, I could attain the riches necessary to please Icza's family, but the Gentile-Jew barrier to our love was as permanent as death itself. And the more hopeless Icza and I became, the stronger our love became. I spent all my spare money on flowers and candy. I followed her to balls and picnics and since our affair was considered a puppy attachment, we were not disturbed by her family.

About fifteen months later, one member of my employer's work force withdrew and built a large plant himself at Budafok, about ten kilometers from Budapest. He invited me to a job with him as a billing clerk. It was a very responsible job in the particular field. Few, if any, seventeen-year-old boys were ever offered the position. I was very proud and happy and was also glad of the chance to put a stop to the hopeless love affair with Icza.

I received a salary of 120 Kronen, far more than any youngster of my age had a right to expect. Since I did my work well, I was hopeful that at the end of the year I would receive a respectable increase. The clerk doing the same work in Losonc had a salary of 210 Kronen.

The Budafok plant was new and the owner was struggling. I failed to receive the expected raise, and I had a swollen head as a consequence of earlier success. I left my job.

CHAPTER 2

America

★

I WENT TO BUDAPEST and started to look for another position but had sad results. Larger firms were unwilling even to consider employing an office worker who did not graduate from Gymnasium, Real School, or at least from the Academy of Commerce. Lack of German disqualified me immediately. It was hardly possible in Hungary in those days to be engaged in commerce without a knowledge of the German language. I went around answering employment ads for weeks and weeks and advertising myself as long as I had money to pay for ads. But no results.

I could not fall back on the family, for my mother was poor. She was hardly able to manage with Charlotte and Regina, and my brothers justly felt that I did not have any business leaving my former position for it was far better than the position that Géza had with his education.

I roomed with four of my ex-schoolmates of the Normal School who were about to graduate within a few months, and I was eating mostly at the People's Kitchen. Every now and then, I managed to earn a few Fillers by addressing envelopes or by seasonal employment lasting a few days, but I went hungry quite often during those months. I went to Jakab every evening to keep his books, and occasionally he treated me to a supper, but I did not tell him how desperately I lived. When I was very hungry, or when I did not have the 4 Kronen for the monthly room rent, I sold a suit of clothes or a pair of shoes at the second-hand market.

Manual labor was out of the question. Skilled workers have to serve three or four years of apprenticeship, and common laborers have to have a book proving their honesty and entitling them to accept such labor work.

And requesting such book, or accepting work as a common laborer, would have de-classed me forever with no hope of my ever crossing the class line again. The caste system was strong in Hungary, and if I had offered my services as a laborer, I would have become an outcast. When in my desperation I did go out a few nights to shovel snow, I took good care to turn up my collar and to hide my face from passers-by.

I remember a winter day when I had only 8 Fillers and purchased eight roasted chestnuts on a street corner. That was all the food I had on Christmas, 1906. I was a great fellow for hindsight, and the knowledge that all this was due to my own fault did not help me any.

One day in February, 1907, I was playing chess with one of my roommates who told me that his parents left in the previous year for America. And, as soon as he graduated, he would join them. From then on, the chess game did not interest me. I had found a solution to all of my problems. For besides being jobless and penniless, I had another, far graver problem — the worry about the military service.

After reaching twenty years of age, every Hungarian boy was conscripted for the Austro-Hungarian army for three years service unless he had the proper education for the so-called "volunteer" year, which is not a volunteer year at all. Graduates of Gymnasium Real, Academy of Commerce, or Teacher's Normal were drafted for volunteer year and upon completion, they became officer's candidates, a rank somewhat similar to America's Warrant Officer, and they immediately were placed in the reserves without further service. Since I failed to graduate, I was in for three years service, a brutal servitude in those days with enlisted men being subjected to corporal punishments at the hands of any sadistic peasant corporal or sergeant.

America was the answer for everything. America meant escape from my impossible situation, and it meant the hope of making a few dollars with which one could buy a diploma from a private Academy of Commerce, thus entitling one to the volunteer service. I was a little over eighteen years old and, within two years, I should be able to return to Hungary with at least $100.00.

Of course, America was an impossible dream. The first requirement for going to America was a passport. In 1907, one was free to travel anywhere in the world without a passport, except in Tsarist Russia[1] and if one were an emigrant from Hungary. The Hungarian government, controlled as it was by large land owners, was reluctant to permit cheap labor to escape the country, and prospective emigrants often bunched up into religious processions going to an Austrian shrine and thus crossed the Hungarian border. Since I was not of Hungarian adult age, mother's consent was needed for any application for passport. This also meant that a family council would decide my fate.

The cost of steerage ticket was about 200 Kronen. Railroad fare and other expenses would come to about 50 Kronen. But I had heard that one may work one's self across the ocean, and I knew that I had to find the means of going to the fantastic new country. The same evening, I went to the library, found and read two books about America, and worked out a scheme whereby I would obtain my mother's consent.

Next day, I sold my last treasure of more prosperous days, my watch, for 12 Kronen, and I was on my way to Fiume, the then Hungarian seaport. I paid 10 Kronen for the railroad ticket. It was a twenty-four hour trip, and on the train I became acquainted with a few construction workers who went to Fiume to build a large government building. By spending 60 Fillers for bread and cottage cheese during the day's travel, I arrived in Fiume with a little over 1 Kronen. Upon ascertaining the possibility of employment on a sea-going steamer, I wired home that I was on my way to America and requested Mother's blessings in care of the telegraph office.

When I failed to receive an answer to my telegram by evening, I went to my new friends, the construction workers, who fed me and put me up for the sleepless night. I decided that if my scheme failed I would apply for manual labor for I was fed up with going unfed week after week and having no prospect of a suitable job.

Next morning, I knew that my scheme worked. There was a wire and 15 Kronen requesting me to return home, at least to say good-bye.

The family council was waiting for me in Mátraverebély. Without exception, my sisters, brothers, and in-laws were present, and they took me in hand immediately. They pointed out, in very forcible language, that I had caused the family plenty of humiliations by having been expelled from school, and now I was ready to put the crowning shame on them by this insane idea of going to America.

Before the First World War and during the Golden Age[2] of the middle class of Europe, only peasants, proletars,[3] and criminals migrated from

Eastern Europe to the United States. My family reminded me that they would have to stay home and face relatives and friends who would never believe that I "did not do something" before I escaped overseas.

The family council had lectured my mother and pointed out that as soon as I went to a Budapest school, away from the family control, I went bad. The family said that I left a perfectly good job because I did not want to work. They also said if Mother let me go overseas, I would certainly go to the dogs.

Mother was the last to say her say. To my great amazement, she sided with me and reminded my sisters and brothers that when I was expelled from school and needed encouragement and love, almost all of them turned against me. And with the exception of Géza, none of them had a good word to say; therefore, they lost their rights to decide my fate.

Mother further said that in his younger days, my father—for whom she had the greatest love and a respect bordering on reverence—also dreamed about America, and my wish might be the continuation of his dream. And to the astonishment of the family, she announced that she was ready to give her consent.

The village Notary also tried to talk my mother out of her decision. No one else had left Mátraverebély before, or since, for America, and he was proud of this record that was pleasing to his superiors.

While waiting for the passport, I wrote a letter, on mother's advice, to a by me unknown cousin in Brooklyn who left Hungary about fifteen years before after a business failure. I asked my cousin to help me with advice upon my arrival. Within about four weeks, both the passport and the answer from Brooklyn came. I was very much impressed with the letterhead which showed that Mr. Géza Weiss was a "Manufacturer of Soap." The letter stated that my cousin would wait for me at the steamer.

The date of my departure was set. The family was together again—sisters, brothers, in-laws, offspring, and I was subjected to a final, last-minute siege to remain home. Mother was baking poppy seed biscuits for superstition held that such biscuits would bring back the departing member of the family. As Mother was bending over the dough, her tears were dropping into it.

Finally, I said good-bye to neighbors and friends, and each of my brothers and sisters, except Jakab, the cabinet maker of Budapest, slipped a few Kronen into my hand as I kissed them. Jakab was still unreconciled, so much so that while we had to travel on the same train as far as Budapest, he refused to take the same coach that I did or to speak to me.

I was happy, very happy, and I only had lumps once in my throat when my mother kissed me good-bye at the station and whispered into my ear that she never would see me again. Luckily, she had several other occasions to repeat the same farewell in later years.

I had over 30 Kronen when I arrived in Fiume, and I was told that the *SS Carpathia*[4] of the Cunard Line would be there within five days. There were several other youngsters in similar situations, and we enjoyed the wonderful spring days on the Adriatic seaport. We took excursions to the famous winter resort Abbázia,[5] and we got acquainted with some girls, many of whom were employed in a cigar factory.

The night before the steamer was due, we did not go to sleep. We were sitting at the dock when the—to me—tremendous bulk of the *SS Carpathia* appeared on the horizon at 3:00 A.M. She did not dock until 5:00 A.M., and the sea-going ship was such a tremendous sight that in my excitement, I did not even think of my job. Unloading, loading, etc., started immediately, and I was amazed at the big loads handled by the great derricks.

At 9:00 A.M., we were permitted to enter the Purser's Office,[6] I stood there in a line of about forty applicants, and I knew well that only a few would be hired. With a sinking heart, I noticed that only husky young men were selected, and when Mr. Strimpler took my passport, I had sudden tears in my eyes. The passport said that I was a clerk, plus I was a skinny boy. Neither of these qualities was a very good recommendation for heavy work. Perhaps the tears or perhaps something else moved the Purser to take me on. Two hours later, I passed the physical examination.

I was signed up as a steward, and the *Hungarian-English Interpreter and Dictionary* I purchased in Budapest told me that steward means "waiter." I was to receive four Pounds per month—the equivalent of $16.00 or 80 Kronen. The other youngsters and I also were instructed to be on board with all our belongings at 5:00 A.M. the next day.

This last day was one of the happiest I spent in Europe. I wired mother the date of my departure on the *SS Carpathia*, cabled my cousin in Brooklyn, and for the first time since I left Losonc, I wrote a letter to Icza. I told her that I was leaving for America with the great hopes that I would find my fortune there and that perhaps we would see each other again with no barrier between us.

And since I still had about 6 Kronen left, I treated two of my colleagues to tickets to the theatre, and we enjoyed an excellent performance of *The Merry Widow*.[7] The performance was so moving that the whole audience sang the famous waltz.

We still had a few Fillers, and after the show we went to the Café d le Villa and listened to gypsy music until closing time. The remaining 8 Fillers went to an old woman beggar. I boarded ship without a Filler in my pockets but with the prospect of receiving approximately $10.00 in New York. The trip was to take about twenty days.

Waiting at tables did not seem hard, and to do it in style, I put on my best (of two) blue cheviot coats with a pair of white flannel pants. The foreman told me in terse voice to take a pail of water, some soap, and scrub the stairs leading from the kitchen to the steerage dining room. And when with somewhat deflated ego, I tried scrubbing by bending down, he promised me a few well-aimed kicks to the place where it would do most good unless I kneeled down and worked properly, like the rest of the boys. And the language he used would put to shame any GI language.

I had no choice in the matter, took off my coat, and ruined my white flannel pants. I told myself that the pants would be the first thing I replaced in America. The same afternoon we received a blue denim overall and a uniform cap.

Next morning, about two thousand passengers came aboard, and at 1:00 P.M., I was standing on the deck when the steamer left. I do not know whether I was happy or not, for with the last glance at the Hungarian city, I knew that I had kicked out whatever ground there was under my feet and burned all bridges behind me.

Contrary to all the steamship company leaflets that showed pictures of nice cabins and recreation rooms, eight tenths of the steerage passengers slept in huge dormitories, each housing about three hundred passengers in double bunkers. Our stewards' days started by carrying food two flights down and dishing out same to the passengers who were seated at long tables, about thirty passengers at each table. My buddy, Eugene Szabo, and I had to take care of three tables and two separate bunches of passengers for there were two sittings. Our own breakfast came after this with plentiful food for both passengers and crew, and breakfast was followed by washing up the dormitory. And with about three hundred passengers, some seasick from the first to the last day, there was plenty to wash.

At 11:00 A.M. daily, the Captain went thru the ship with a searchlight in his hands to note the smallest dirt, if there were any. When inspection ended, lunch was served. Lunch was followed by washing partitions until about 4:00 P.M. when we were lined up on deck and roll was called. A few minutes were spent on deck before we started serving dinner, after which we had to scrub and wash the dormitories once more. This scrubbing of the dormitories was the most unpleasant job of all. From about 8:00 P.M. to 11:00 P.M., we had to carry sacks of potatoes from the hold to the kitchen, and when the carrying was finished, we were finished with our day's work.

The stewards worked for the 4 Pounds, but I did not mind that as much as I did the fact that we were not permitted to leave the ship at Naples,[8] Palermo,[9] and Gibraltar,[10] although we stopped at each of these ports for about one day to receive cargo and a few passengers. Nor did we have time to spend on deck to at least look at the romantic places.

On the ninth day when we left the Mediterranean, a heavy storm was encountered, and I became sick, terribly sick. I am still subjected to the terrible sickness when on the sea, even after many crossings and several long trips.

But sick or not, I had to do my chores. Once, when carrying a high stack of plates to the kitchen, the steamer suddenly went out from under my feet, and I slid down and broke about forty plates. I was surprised that nothing was said by the foreman, but I did not know how many surprises were in store for me later.

Another seasick incident occurred when I was distributing food at the head of my table. I became violently and actively sick, and each of the thirty eating passengers joined me. It was an interesting, if not pleasant, sight, and I felt it had served the passengers right for they had no business to eat in storms when their betters all were seasick.

I was sick for about four days, but aside from that, we had a jolly time, notwithstanding the long hours that we had to work. We were young and about to reach the Promised Land, and each of us had our own separate plans and dreams. I was hopeful that I would not land with empty pockets for it was rumored that passengers usually tipped waiters at the last meal.

Evidently, Eastern European peasants were not aware of the tipping custom, or else they were just naturally stingy, for all we collected from close to two thousand passengers was $.60. Knowing that I would receive a pay of about $10.00 upon debarking, and that my buddy would remain on the ship for one more trip, I let him have the whole collection. After all, what is $.30 to a man who will have dollars tomorrow and whose cousin, a manufacturer, will wait for him at the steamer?

On May 7, 1907, we ran up the deck in great excitement before serving breakfast. Land was sighted! On that beautiful May morning, the rising sun shined down on the green fields of Sandy Hook with the Statue of Liberty and the skyscrapers of Battery Place in the distance.

God was good to me for I can't say that the thrill of seeing America from the deck of a steamer came to me only once in my lifetime. I was granted another such thrill thirty-seven years later, but either of the two would have been enough to keep me in courage to my dying day. And that first one glorious look was and is cherished by me all the more because it was really only one look. I was reminded of Moses with one look to the Promised Land when we were chased down to serve breakfast.

Three hours later, the *SS Carpathia* was in Cunard's dock. We learned then to our dismay that there were three other ships loaded with immigrants ahead of us, and thus we would have to wait for our turn to unload — about two days. Approximately twenty thousand immigrants were delivered weekly in those days to America.

We waited two long days in dock, and workers were not permitted to leave the dock for the steamship companies were fined for any sailor who skipped without going through Ellis Island.[11] Those of us who were ready to leave the ship were told that we would be taken to Ellis Island even before the passengers and would be paid off after our admittance.

My cousin did not show up, and I was somewhat worried about the requirement that each immigrant had to show $10.00 at Ellis Island. However, I was hopeful that my cousin must have learned about the delay and would contact me at the Island.

Finally, on May 9, 1907, we were taken to the Island of Sighs. No passport or visa was needed in those happy, free days. The doors of America were wide open to anyone willing to work. Immigrants were inspected only for the dreaded trachoma[12] and were turned back only if they were infected with that Egyptian disease.

When I was facing the officer whose business was to talk about my financial situation and whether I had the required $10.00 or not, I felt so offended by his question that I forgot to worry about the lack of money. He addressed me in German and in terms of "thee" by asking: "Wie viel geld hast du?"[13]

I told him as best as I could that I expected to be paid and would have over $10.00. I was passed.

While waiting for the ferryboat to take the *SS Carpathia's* servants ashore, I borrowed $.03 from one of my steward colleagues to buy an apple for lunch. It was about 3:00 P.M. when we were taken to the British Consul's Office to get paid off, and my cousin still did not show up.

CHAPTER 3

Investment in America

★

A RUDE AWAKENING WAS waiting for me with the pay-off. We stewards were told, first of all, that the $2.00 head tax was paid for each of us for each immigrant had to pay that. Not one of us was aware of this because, for passengers, the head tax was included in the price of the steamship ticket. Some money was deducted for the overall, some for the cap, some for breakage of dishes, and my balance to be received was $.12, upon the receipt of which I had to sign that I was paid. I paid back the $.03 I borrowed, and I was standing there with $.09 to my name—the total capital I invested in my life in America.

I was a stranger in a strange world, with only $.09, no manufacturer cousin to take care of me, no ability to speak English, and nowhere to turn for shelter or my next, now overdue, meal. I was standing there dizzily until one of my colleagues made the remark that he would gladly pay $.25 to someone for carrying his small wooden trunk to the Rakoczi Hall. With the exception of myself, who was all set to go with the cousin, all the boys planned to go to the Rakoczi Hall, a Hungarian saloon on East Fourth Street. We had secured a card in Fiume that showed the address of Rakoczi Hall and advertised rooms for $.25 and meals for $.15.

I quickly took my colleague up on his offer to pay the $.25 and, leaving my two suitcases behind, I carried his trunk on my shoulder from the Battery Place to East Fourth Street, a distance of six miles. Of course, the elevated train would have taken us there for $.05, with trunks and all, but how could a greenhorn know about such things?

It was about 8:00 P.M. when I arrived with my own suitcases on my second trip to Rakoczi Hall, but I had earned my first money in America, and I had the $.25 to pay for a room in the Hall. By the time I had made my second trip, my colleagues wised me up to the wonderful institution of the free lunch, one of the best American traditions that has gone with Prohibition. Until Prohibition, there was a long table in every saloon in America, a table loaded with bountiful cold cuts, pretzels, and crackers, and any customer—even if he spent only $.05 for a schooner[1] of beer—was welcome to eat as much as he wished for free.

Later, I became acquainted with saloons where a plate of hot soup was also given at midday to boot the cold cuts and with other saloons where hot frankfurters were available. But, on the current occasion, the table of the Rakoczi Hall looked good enough for me. Upon buying the beer that I did not like, I gorged myself on cold cuts to the bursting point. And still I had $.04 left with which to greet my first morning in America.

That night, I decided that I would not seek out my cousin. I had managed to master the worst situation of my life. I had stood on my own two feet in the face of insurmountable obstacles; I would go my own way further. One can be very proud when one is eighteen years old.

We woke to a beautiful spring morning and after having a cup of coffee for $.02, and two rolls for another $.02, I was broke once more, but in good spirits. We went shopping to look up labor agencies in the East Side, all of which were looking for porters, bus boys, timber men, and coal miners. Since I was warned by my brothers that immigrant ex-army officers and such other educated trash who had escaped the prosecuting attorney usually wound up as dish-washers in America, I chose the toughest job offered—coal mining. Free transportation was promised, and I had talked four of my colleagues into joining me. They were in better circumstances for all of them had a few dollars, but I had to take something offering immediate shelter and food.

We signed up after some persuasion of the labor agents and were promised shipment on the same afternoon. Mr. Engel, the labor agent manager, cinched down the bargain by treating all of us to a regular meal at $.15 a throw. At 4:00 P.M., we were joined by a few other Hungarians at the agency, including a Hungarian woman who had arrived the same time from the Old Country and whose husband took her down to the coal camp to keep a boarding house for Hungarian miners. Carrying a big sack full of bread and lots of cold cuts, we left New York at 5:00 P.M. for Thacker Mines,[2] West Virginia.

The train with its plush seats was a new testimony to the inconceivable riches of America. In Hungary, middle and lower-class folks travel in third class with wooden benches, the second class is somewhat better, but only seats in the first class are covered by plush. I was afraid for hours that Mr. Engel would be fined by the conductor for putting his shoes on the opposite seat. Ice water on the train was another wonder, for in Europe one buys water at larger stations from kids or from older women selling water by the glassful.

As we rushed through New Jersey, Pennsylvania, Maryland, and Virginia, each new farm and each building was a revelation, as was the men working in the fields not in peasant costume, but in regular store clothes. For that matter, I was much surprised to see working men in dirty clothes coming into the coach and sitting right down on the nice plush seats. I was somewhat taken aback by so many frame buildings and could not see how they did escape fire. I knew that I never could write home and tell family and friends that peasants were working in pinch hats and patent leather shoes in the fields, nor would family and friends ever believe me if I wrote about the free lunch.

Of course, I did a lot of day dreaming on the long trip. My colleagues and I were told that the average earnings of a coal digger were $3.00 a day, and counting twenty-five working days a month, I was to make $75.00 a month, all of 375 Kronen, the salary of a top man in any large office in Hungary. On Mr. Engel's advice and to encourage the man with his newly-arrived wife, we agreed to board with the man and his wife, and we were to pay $16.00 per month. It was still to leave me enough to save—say—about $50.00 monthly, more than Géza's salary was.

Of course, I also worried about the little matter of mining coal. I remembered several books I had read about the dangers, the explosions, the gas, etc., and the descriptions about the miners bending or lying on their bellies while digging coal. I tried to chase away these worries with considerable success by reminding myself of my experience at the Consul's office and patting myself on the shoulder by manly by-passing my cousin.

Next day, about 10:00 A.M., we had to get off the train at Bluefield, West Virginia, to wait four hours for another train. After exploring the main street alongside the railroad, we were treated by Mr. Engel to coffee with our cold lunch and ice cream after lunch. It was unbelievable! Ice cream for prospective coal miners. Back in Hungary, ice cream was known by very few city people and tasted only by the wealthiest of them.

It was about 6:00 P.M. when the train pulled into Thacker, West Virginia, and two wagons were waiting for us to convey us to Thacker Mines, four miles away. Those were the first mule teams I ever saw—an additional wonder.

The scenery was beautiful. The wild mountains of West Virginia reminded me of the Mátra Hills surrounding Mátraverebély, and everything was green around us. Since it was getting dark, our drivers had miners' caps on their heads, and the torch-like lamps increased the romantic setting. And yet, I had a hollow feeling in my stomach as we were nearing our goal and our new lives. The hollow feeling lasted only until we reached our destination.

We were unloaded at the company's boarding house and were immediately ushered into the dining room. Wonder of wonders! Tables with white tablecloths, snow white bread in tall piles, large plates full of meat, potatoes, as well as other vegetables, plus butter, jelly, coffee, and milk on the table. Why, millionaires could not live better in Hungary. White bread is unknown on a poor Hungarian man's table, butter is eaten only if the supper consists of bread and butter, jelly if the supper is bread and jelly, and coffee is used only for breakfast, but not by coal miners.

Why worry then about the mining of coal? If miners could afford such standards, it must be a fine occupation, indeed. And as to wonders of the *Thousand and One Nights*[3] unfolding, the dinner did not finish it.

We were assigned to beds with real spring mattresses. I don't believe that there were ten thousand spring mattresses in the whole of Hungary. Straw mattresses were used in Hungry, except by the very wealthy people. This may be monotonous to read, but the breakfast was the real thrill for us—ham and eggs, hot cakes, butter, jelly, syrup, and coffee. How could I ever tell tales like this upon my eventual return to Hungary? A plate of sourdough, or bread and raw bacon at best, is the breakfast of the worker in Hungary. After breakfast, we were taken to the company's store for the necessary outfit before entering the mine.

Overalls, wide-toed brogans, miner's cap, gloves, shovel, dinner bucket, and oil can were furnished to each of us, and we were shown the intricate art of filling the lamp on our caps with burner and oil. Nowadays, electric lamps are used by the miners and, while the miners do not like to carry the necessary batteries, they now do not have hot oil dropping occasionally on their hands as in the days of old. We were told in the office that cost of the "free" transportation would be deducted from our earnings, and our total indebtedness for equipment and transportation was $32.00 per person.

Each of us received $2.00 worth of scrip[4] which served for money in the company's store, and we were told that we could draw scrip any day against our earnings. This scrip business was a wonderful racket in the coal camps. The camps paid once a month, and if a miner needed cash between pay days, he had to sell his scrip, usually to the superintendent or foreman at a discount of 25-30 percent. And, of course, the miner was forced to buy everything from the company's store where scrip was accepted at face value.

As soon as my colleagues and I were "dressed in," we were taken to the mine which was about four hundred feet up the mountain. I was pleasantly surprised by the height of the coal which was five feet, and that meant that we were not to work on our bellies. As we entered the dark drift mouth,[5] I again experienced that hollow feeling. But I had to resist the urge to turn back. One hundred feet further into the mine, I would not have been able to return alone, anyhow.

While I was scared good and proper, there were some romantic aspects to the first trip into a coal mine. At intervals, loaded trucks of coal pulled by small mules and driven by Negroes went by. The Negroes wore torch-like lamps on their heads, and the caps were just as the new mining cap given to me. But, all the romantic sensations and all the romance went out of my life when we turned into a working place called a "room," and I was told by the foreman that the room would be mine and that I would work there—alone.

There was a big pile of loose coal shot down during the night shift and an empty car at the end of the truck. I was given ten brass checks with my number, and one check was to be put on each car loaded by me in order for me to receive credit for same. Before the foreman left with the other new miners, he loaded one third of a car full of coal in about ten minutes, thus demonstrating to me the ease with which the coal miners worked. He told me that the driver would come around at regular intervals to pull the loaded cars, each holding about four tons of coal, and to place new, empty cars in my room. The foreman said that the shift would be over when the room was cleaned of loose coal. But, he also said that since it was late and that since I was an inexperienced man, it would be quite all right for me to load about three or four cars, or approximately half of the loose coal. We received $.50 for each loaded car.

I was frantic when they left me, and I realized that from then forward, I would be strictly on my own. I started loading, and it was more difficult than I expected. It seemed that in the hands of the foreman the shovel slid easily on the smooth floor, but in my hands, the shovel was against a hundred bumps as I pushed it against a thousand big pieces of coal. I closed my fists on the handle as hard as I could, a fact which caused immediate blisters, and I often pushed my hands with my thigh. And when I did have a shovel full of coal, the shovel was heavy, and I had to lift it high to throw the load into the bank car. I tried working in gloves but somehow I never could get used to them, and throughout my years of manual labor, I never mastered the art of working in gloves. Soon my hands were full of burning blisters, and my eyes full of tears, plenty of tears.

I was in terror lest the driver would come for the load, and I started throwing big lumps by hand, not realizing that the edges of the lumps were sharper than any razor until my hands were cut to pieces and were bleeding from hundreds of cuts. After an eternity, which might have been at least ten minutes, the driver did come. The car was not loaded, and he left with a wide smile on his black face. By that time, my face was also black with coal dust smeared by my tears. My lack of progress meant one less load to pull and one less empty to place for the driver, and he was pleased. He told the tale of the crying miner in the next room, and within a few minutes, my neighbor came to visit me.

I never will forget the smiling face of young John Hudak, a Slovak boy about my age, nor his kindness. He spoke in Hungarian, and instead of laughing at my tears, he assured me that I would soon get the hand of the art of loading coal. He said that loading was hard at first, even for him, accustomed as he was to hard work from childhood. He also said that the blisters would hurt for only about two weeks, then would develop into calluses, and the cuts would heal if I urinated into my hands often enough. And to boot the good advice, he helped me load the first car of coal; in fact, he did the loading, and I the helping.

And John Hudak kept on, encouraging me with all the eloquence at his disposal and promising to visit me often and to help me until I became a full-fledged coal miner. Indeed, he came to me at least twice a day during my short stay at the Thacker Mines, and he never failed to load at least half a car at each visit. All through my forty years of life since then, one of my deep regrets is the fact that I left the place without saying good-bye to John and thus lost one of my first friends in America, the first of the many who extended me a helping hand. I was unable to trace John in later years.

The driver was surprised to see a loaded car on his trip to my room on the following day. He dropped the car to the entry and invited me to help push the empty into my room. The tracks were laid in favor of the loads, and empties had to be pushed up the face by manpower.

It took me what seemed to be years to load the second car. The driver had to pass me twice before it was loaded, but finally I had accomplished the task alone. It was not loaded properly for I did not know how to build up the hump with large lumps. But, I was considered loaded, anyhow.

I was fearful that it must be at least 5:00-6:00 P.M. and that I would be left alone in the mine and certainly miss supper. Leaving the loaded car behind, I left for the day, praying that the tracks would lead me to the drift mouth. Reaching the outside world, I was surprised to see the sun high up and realized that it could not have been much later than 1:00 P.M.

I was tired, terribly tired. Every bone in my body was aching and my hands were burning, but it was much easier going downhill than up. It was a beautiful summer day. I came through the ordeal in one piece, I had made $1.00 by loading two cars of coal, more than enough to pay for my board for the day, and I was eighteen years old. I picked wild flowers on the mountainside and arrived home with a be-flowered cap on my head.

As I approached the boarding house, my face was red, but I soon discovered that I was not the first one to quit. My buddy, Dezso Marosi, was at home and already washed. And he loaded only one car. There was a shanty to each miner's house to wash and change. There was a washtub, plenty of hot water, and soap. The rule of life was that the Mrs. had to wash the back of the first miner home, and the miners washed each other's backs after the first was clean.

The Lord knows that my colleagues and I worked hard on the *SS Carpathia*, but it was child's play to that first day in the coal mine. After a little rest, we went to the company's store where, about 5:00 P.M., the load sheet was posted, and the brass checks taken from the loaded cars were hung. Noting that the average miner had from five to seven checks, meaning earnings between $2.50—$3.50, we took off our own checks with little pride. As we met several other Hungarian miners, we were subjected to coarse jokes and razzing. Most of the other miners asked us how long that we were going to stay, recounted the history of previous transports, and remarked that nine tenths of the "transportation boys" usually skipped within a very few days.

Silly talk, it seemed to me then, as if one could skip and leave heavy indebtedness behind when one did not know where to go and could not even speak the English language. Silly talk, as if one could not see the well-armed mine guards parading around the porch of the store with a heavy .45 on their belts, and as if one would be willing to start one's life in America in the penitentiary!

There were a few veteran miners who did not razz us and from them we learned a very disturbing fact. Payday was once a month, and one month's pay always remained with the company until a miner quit. It did not take me long to figure out that I would not see the color of money for a long, long time. For even if I mastered the art of loading and made—say $50.00-$60.00 a month—payment to me would not be due until two months later, at which time I would owe over $32.00 to the company and another $32.00 for two months' board, not to mention the scrip I might draw for necessities. I would therefore not see any money until the end of three months, at best, and I did not have a single penny to my name.

Hearing the payday news didn't surprise me, and I was quite willing to believe that city-bred "transportation boys" skipped soon after learning these sad tidings. It was also explained by some miners that skipping was not so hard and that boarding bosses—as owners of boarding houses were being called—did not mind their boarders' skipping. In fact, the bosses were encouraging the boarders to do so. The board payment was guaranteed by the company, and runaways were forced to leave behind most of their belongings, clean velvet to the boarding boss.

I worked eight days in the Thacker Mines; eight hard, hopeless days. I became convinced that I would not receive any money for five or six months, for on my best days, I loaded only three cars of coal. My friends of the *SS Carpathia* did not do as well. By the end of the eighth day, even Hudak was convinced that I never would become a good coal miner and advised me to seek outside labor. He told me that before he came to Thacker, he was working on docks in Norfolk where $2.00 was paid for a ten-hour shift, and the work was much easier. After much discussion and many conferences, the five of us *SS Carpathia* men decided to skip.

On Monday, we told the boarding boss at supper that we would return after supper to clean up our places, a practice that was permitted. He knew darned well that we did not become so industrious suddenly, and he said that it was a good scheme.

Putting on three pairs of underwear, three shirts, and stuffing several handkerchiefs into our brogans, I took a last look at my suitcases and, with heavy heart, I left my belongings. And with dinner buckets in our hands, we climbed toward the drift mouth for the last time.

Hiding in the bushes, we waited for darkness to fall. Then down to valleys and up to mountains we went toward the main railroad line and toward the town of Thacker. We always were careful to keep a respectable distance between ourselves and the railroad track. On reaching the main line, we started to walk east alongside of the track until daybreak. At the break of day, we left the tracks and went to sleep in the woods. Our dinner buckets came in handy; we were hungry.

We slept during the day and started up again on the railroad tracks in the evening. The next morning, we went to a small store and purchased bread and thick, salted bacon. We went to sleep in the woods once more.

The third morning, I found a Norfolk & Western Railway Time Table at the small depot of Glen Alum,[6] West Virginia, and when looking over the map, I realized that we were making a great mistake by going east. I saw on the map that all of the great industrial cities I knew about were west of us, while there was nothing east except Norfolk. Columbus, Pittsburgh, Cleveland, Toledo, Detroit, Cincinnati—I knew something about all of them and was especially intrigued by Portsmouth.[7] I remembered that the Russo-Japanese Treaty[8] was made at Portsmouth and thought that it must be an important city not very far from where we were. It took me some time to learn that the Russo-Japanese treaty was made in another Portsmouth and that there were several towns and cities in the United States with the same name. By that time, I was the leader of the gang, the youngest one, and it did not take much to convince my buddies that we had to turn back and seek our futures in the larger cities.

By walking nights only, it took us about ten additional days to reach Portsmouth, Ohio, and we were a sorry sight when we got there. At times, we washed in some creeks, but we were filthy, nevertheless, and had not shaved for two weeks. But, we were emboldened by our success in avoiding arrest and washed, shaved, and put on clean shirts at the train station. We also shed our extra underwear and shirts and tied up our possessions in handkerchiefs. About 6:00 A.M., we went to a lunchroom for our first hot breakfast within two weeks and then started to explore Portsmouth.

CHAPTER 4

I Became an American

★

WHAT FOLLOWS NOW is a fairy tale certainly worth telling: A little after 8:00 A.M., my *SS Carpathia* colleagues and I reached the Portsmouth City Hall, and I knew what it was. Upon my urging, we went in, and like people who have little or nothing to lose, we stepped into the first open door. I went to the man sitting there at a desk, and pulled out my little green book, the *English-Hungarian Interpreter and Dictionary*. I pointed to the sentence that said, "I want work." I had no idea what the amazed clerk answered, but I pointed out another sentence stating that, "I speak German." Literally, it was not true, for I did not speak German, but I was able to make myself understandable in German and two of my buddies did speak German fluently.

When the bewildered man sufficiently recovered himself, he went into another room and returned with three other men within a few minutes. For awhile, the men tried to talk to us, but our answer was repeatedly a referral to my green book and our pointing out that we wanted to work and that we spoke German. Finally, the office workers motioned us to sit down on a bench in the hall, and we saw that one of them telephoned somewhere. We did not know whether the call was to the jail or to the requested work. We spent an uneasy fifteen minutes on that bench. In Hungary, vagrants or strangers without any visible means of support are usually sent home under police escort, and the hometown is assessed for the expenses which, in turn, are collected from the vagrant's family.

Within about fifteen minutes, a jovial-looking fat man came in, and we were called into the office once more. The man spoke German far better than I did, and our ready-made story was slowly told to and through him.

Our story was that we landed in New York five weeks ago and could not find jobs there, so we started walking and looking for jobs everywhere, but could not find jobs because we could not understand, much less speak, English. Our story included that we had no money but were eager to work at anything, for any pay.

It took about half an hour to have our story told and interpreted, and this was followed by a conference between the Americans. Suddenly, our interpreter asked us if we were willing to work at street paving. I told him that we came to America to work and would be happy to do anything. The man then said that we would start working Monday and would receive $1.50 for ten hours of work. It was Thursday, and I had to tell him that we had no place to stay and no money to carry us until Monday.

Following another conference between themselves, one of the men called somebody on the phone once more. At the conclusion of the phone conversation, our German friend said that a man would take us to a place where we could board and pay $3.50 a week for same.

The boarding house was about four blocks from the city hall. It was a clean place, and its owner assigned two rooms to the five of us. Even my present, still poor English, is inadequate to describe my happiness. It is true that coal miners' earnings were a lot more, but with $9.00 a week, and with paying $3.50 for room and board, I would still have at least $5.00, the equivalent of 25 Kronen, saved every week.

Of course, the boarding house provided new American wonders. At lunch, we pleasantly were surprised at the variety of food placed before us in miniature dishes instead of one large plate as is the custom in Hungary. But, we found the food tasteless. We missed the customary shortening on the vegetables and in the soup, but the meat and cake, as well as the coffee, were appreciated. The white bread did not surprise us; one gets used to luxuries quickly. After lunch, we went to take in the town. As we were sitting on a bench in the small park, I had a smart idea, or a brainstorm. I pointed out to my colleagues that, after all, we were hired by the city as a matter of charity—which was not the case at all—and that now that we had shelter and food, we ought to try to stand on our own feet. And I thus decided to look for work elsewhere.

My colleague, John Szalay, and I started to look around and tried to approach two factories, but we were not even permitted to enter. The third place we approached was a lumberyard and, with my green book, I went to the man in charge and pointed to the well-worn sentence stating that I wanted work. After some difficulties, we were hired, and the foreman pointed to 7 on his watch to indicate that we would be expected to be ready for work at 7:00 A.M. the next morning. By now, people were used to immigrants who were unable to speak English for we were entering America weekly by the thousands.

I felt like any self-made man, rushed to the city hall, and requested them to call our interpreter for an important session. When the interpreter came, I explained through him that I had other work and also my reasons for getting work elsewhere. I told him frankly that my colleagues and I appreciated the assistance of City Hall, but we would like to relieve them of the burden. Sooner or later, all of us would find work somewhere, and three of us would still show up Monday for paving.

The kind people at the Portsmouth City Hall were not offended. On the contrary, they seemed to have liked the spirit. One of them patted me on the shoulder and assured me smilingly that it was all right. Our interpreter invited us to go with him. He was a saloonkeeper and treated each of the two of us to a glass of beer. Then, John and I rushed to our three other buddies to break the cheerful news.

Unfortunately, the next day turned out to be somewhat less cheerful. John Szalay and I did report for work and carried stacks of lumber hither and yon, but we were unable to understand the orders. The foreman had to be ever at our heels, and by noon, he got tired of the system. We were fired and received $.75 for the half day's work.

With considerable deflated spirit, John and I went back to our saloonkeeper and requested him to break the news at the city hall gently and to ask permission for all of us to report for work on Monday. He did this by phone and reassured us that it was all right and that there were no hard feelings at the city hall. He also treated the now five of us to the customary schooners of beer.

All of us went to the park once more and what follows now makes me blush, for I certainly would not have the nerve to repeat now what we did then. An old German-speaking man started to talk to us in the park and soon learned half of our experiences, the Portsmouth half. The old man was a laborer and lived in another boarding house, called The American House, right on the bank of the Ohio River. According to him, the American House was much better, the food more plentiful, and the charge was only $3.00 per week. He invited us to visit him in the boarding house; in fact, he took all of us there at once and suggested that we move there.

I was aghast at the suggestion and told him that we already were indebted at the other boarding place and could not think of moving. But the old man was persistent and fifteen minutes later, the five of us were on our way to the city hall with the old man in the lead. And he told the man at city hall—I supposed he told them for I did not know—that all of us would like to move because of the difference of $.50 per week, if they had no objection to the move. And to my great relief and joy—for I expected to be kicked out of city hall—the men of city hall seemed to have the time of their lives with us. They called up the owner of our boarding house, told them to let us take our little bundles, and guaranteed that we would pay what we owed at the end of the following week.

We were happy at the prospect of saving $.50 or 2.50 Kronen weekly. And I enjoyed some of my Saturday by sitting on the long porch of the American House and watching the excursion steamers drifting down the Ohio River with more electric bulbs than I ever saw on the King's Palace[1] in Budapest. I decided then and there that with God's help, I also would someday take trips on the luxurious boats. Dreams do not cost anything.

Later that afternoon, the five of us also went window shopping. The ten-cent store reminded me of the story of *Alice in Wonderland*.[2] Clothing was cheap. I saw that with four to five weeks of earnings, we could replace everything we left in Thacker. To celebrate, we bought popcorn and were surprised again at the bounty of a country where butter was wasted even on popcorn.

Monday, we started toward the street indicated by city hall to finally start working. But we did not go there. When we passed a small shoe factory, I was reminded of a book I read in Budapest, the author of which stated that skilled workers in America—such as shoemakers, for instance—were making more money than office workers in Hungary. The book was written by the Secretary of the Hungarian Chamber of Commerce after his visit to the St. Louis Exposition.[3]

I high-pressured my colleagues to come with me to the shoe factory employment office. The young chap there was one of the many Portsmouth residents who spoke German, and I explained to him that we were eager to learn the trade of shoemaking and that we were willing to work as apprentices until we learned the necessary skill.

After speaking to someone in another room, he came back and offered us jobs at $1.00 a day to start with. Some of my colleagues did not like the idea, but I was enthusiastic with the prospect of becoming a shoemaker. Finally, all of us apprenticed ourselves, as we thought, to the trade. The young man promised to call up the city hall after 8:00 A.M. to tell them that we did not skip but were working.

I realized soon that it would be a long time before I would be a shoemaker. I was given a die and a hammer and had to cut small pieces off a cardboard for heels of ladies' shoes. But I was hopeful that later I would receive some instruction and some more complicated tasks.

At lunch, my colleagues and I sat next to the rest of the workers and some picked up conversation with us when they discovered that we spoke German, besides the outlandish language of our own. And we told them the smart thing we did by offering our services at apprentice wages to learn the trade. As a consequence of the intelligence, an excited bunch of shoe factory workers went to the shoemaker's office and, as we soon learned, they served an ultimatum to the shoemaker—a pay of $1.50 for them for the ten-hour shift, as was paid to other workers, or a strike by the workers, or discharge for all five of us. Well, at the regular pay, the shoemaker did not need five men just then, and we were discharged with $.50 to each of us. Downheartedly, we went back to the saloonkeeper, and I did not feel any better by knowing that I was the cause of our inglorious experience.

Our friend in the saloon still received us with a smile, but we did not know what the reaction would be at the city hall. We sat around the phone shamefacedly and bitterly. Then, a message came from city hall. The message was to report for work at street paving without any new scheme or side trips.

We were wheel barrowing bricks on the new job, not an easy work, but not beyond our capacity, and we were chatting once more among ourselves. Suddenly, the foreman came to us and addressed us in—Hungarian. He was a Romanian from Transylvania and had come to the United State ten years before we did. Upon learning our story, he said that we would be a lot better off working in the steel mill about six miles from the town in a settlement now called New Boston. He said that there were about a hundred Hungarians there with four Hungarian boarding houses where we could get Hungarian food.

I told the foreman that after the jumping around that my colleagues and I had done, I would be afraid to tackle anything new, and I did not wish to enrage people in the city hall. I also told him that we would look around at the steel mill and the Hungarian boarding houses the following Sunday. The foreman said that he was satisfied with our work and was not trying to get rid of us, but it would be foolish to suffer on American food when we could help ourselves at once. And giving me $.10 for carfare, he insisted that I go to the settlement at once and assured me that the time would not be deducted from my shift.

When I reached the steel mill, I said that I spoke Hungarian, and one of the laborers was called in to interpret. I was told that my colleagues and I could start work the next day, or the next Monday, and my new Hungarian friend told me that the five of us would get board between the four families keeping boarders.

When the steel mill quitting time came, the foreman went with me to the city hall. He explained to the people there that on his advice, the five of us would work in the steel mill for it was a hardship for us to live in the American boarding house since we could not speak English. He guaranteed that our bills at both boarding houses would be paid.

The man at the city hall phoned the steel mills and verified our new jobs. Then he called up the American House and requested its owner to let us take our little bundles. The city hall man also assured the owner that all of us would pay our debt to the American House within two weeks. And to show that there were no hard feelings, the men at city hall shook hands with me, and one assured me that I would have an interesting career in America. Is it any wonder that I was sold on America?

As my colleagues and I traveled on the streetcar to our new home in New Boston, I went over in my mind the few days I had spent in Portsmouth, and I knew that I never would leave this wonderful country. I compared the police system of Hungary with the helping hands of the city authorities of Portsmouth. Instead of sending vagrant men back home under police escort as Hungarian police did, the Portsmouth authorities went far out of their way to help five bum-looking young strangers, and a businessman left his place of business several times to assist the young bums who really were imposing on his good nature. The work foreman of a sizable group of Hungarian workers would deem it below his dignity to talk to his workers whereas, in America, the foreman gave me money to get a better place to work and then went to bat for me and my colleagues at the city hall. I fell in love with America on that day, an enduring, ever-lasting love that only could be understood by another immigrant. And this love was a happily-ever-after ending to our Portsmouth fairy tale.

When the First World War brought about the first wave of intolerance and alienation to America and Americanism, I was saved of despair by cherishing the memory of those few days at Portsmouth. And even when the Ku Klux Klan[4] was riding rough-shod over the best part of the United States, and when still later the Silver Shirters,[5] the Black Legion,[6] and other similar organizations were preaching intolerance and hatred, I tried to assure myself and other immigrants that the real America was Portsmouth, the real Americans were the Portsmouth folks.

In later years, I often listened to lectures and speeches about America, and I took part in the "I Am An American" celebrations. But, I always felt that the good deeds of five or six men at Portsmouth and the others like them spread Americanism much more effectively than any silver-tongued speaker could. The deeds of the Portsmouth men planted the love of America deep into the breasts of five struggling immigrant boys.

My colleagues and I were distributed between the four boarding houses at New Boston, and I obtained board at Mr. Petre's. He was a young man, his wife barely eighteen years old, and they lived in a large ramshackle building right at the fence of the steel mill's yard. The ground floor was divided into three rooms, one for the kitchen, one for the dining room, and the third was occupied by the family with their two little children. Upstairs, there was only one large room with twelve double beds and little place for anything else. There were twenty-three boarders besides me.

I was somewhat taken aback when I was told that I was to sleep with Joe, and I suppose Joe was not any happier about his new bedfellow. It was the custom—and for all that I know it may still be—to have two men sleep in a bed with each one not knowing whether the other was a victim of some disease or not. But I was grateful for the privilege, anyhow.

I could not help noting that I was received with some justified and some unjustified suspicions by my fellow boarders. I had hardly anything in my possession, and I had no choice but to tell the boarders my whole story, including the escape from the Thacker Mines. My story was not conducive to create confidence; it did quite the contrary. The boarders immediately assumed that a man skipping from one place and brazenly admitting it might skip from another place and not leave empty handed. Therefore, trunks and suitcases were locked right before bedtime. Further, my hands did not show calluses, and white-collared immigrants were ipso facto considered crooks or at least suspicious characters by their fellow countrymen. The other boarders knew that only peasants and proletarians were forced to emigrate in those days. I was careful to talk as little as I had to lest my somewhat more cultivated language give away the fact that I went beyond the compulsory fourth grade in the Old Country.

I did not criticize the beds, which were terrible, and had mountains and valleys with plenty of dirt on the blue gingham[7] "linen." I enjoyed the Hungarian soup, the boiled beef, and the bread served at my first supper in the new boarding home, and I did not show any surprise at the lack of vegetables. Of course, the suspicion of the men hurt me but, with all that, I still was happy as a lark at my decision that I was now in my permanent country.

Next morning, I was assigned to a small group of four men, and we unloaded pig iron[8] from a boxcar. Of course, for many months thereafter, I called the pig iron "bigany," the term that I learned from the other immigrants. I and every other immigrant had to learn English twice—once the Hungarian, Italian, and Polish English, then the American English. Within a few days, overtime was "oberta" and time was "tay" for me.

The rough surface of the pig iron was hard on the hands. Most men were using a leather pad and a few gloves, and I decided to buy gloves on my first pay day. When I did receive the first pay, I walked to town to save the nickel for which I had to work for twenty minutes. When I arrived in town, I decided that forty minutes of work was too much paid for a $.10 pair of gloves. Upon returning home, I made a pair of gloves from flour sacks.

The work was not very hard for a healthy young man of eighteen years; it was child's play compared to shoveling coal into bank cars. My steel mill colleagues worked in a slow tempo as they lifted and carried a pig to the door, and I was careful to lift and carry one on my turn just a little faster than the others. I learned here the difference made if one worked by the ton or by the day—a great difference indeed. Later, we unloaded scrap iron and used and ragged rails, and here again, I did my best to lift my end as fast and as high as the man on the other side. In those happy days, industrial plants were not as well equipped as now and had no electric cranes and other labor-saving devices. Labor was cheap; machinery was expensive.

I was good and tired when the day's shift was over, but I did not find the work beyond my capacity. I had $1.50 to my credit, the most I had made in a day in my new country. And I had the satisfied feeling that I did as well as any other man working with me. A large pail full of beer was purchased every evening after dinner. Though I did not and still do not like beer, I was afraid to refuse my share, fearing that my refusal would offend the bunch. A few days later, when I knew that I had more credit than indebtedness and when I thought that the suspicion of me was decreasing, I told my fellow boarders that I did not like beer. The decision was that since we lived in a co-operative group, or as it was called in "company board," I had to chip in just the same. I was glad that thereafter I did not have to drink, and they were glad to have my share. I know that all this sounds small, but we were counting in Hungarian money, and my two and one-half cents share amounted to 12.50 Fillers, one-fourth of the then daily wage of a peasant in Hungary.

The boarding house food was hard on me for it consisted of meat, bread, soup, and black coffee with no vegetables at all. Without any deviation, we consumed close to a pound of pork chops, bread, and black coffee for breakfast, roast veal, bread, and black coffee for lunch, soup, beef, bread, and black coffee for supper. The boarders were satisfied with the fare. They, as other Eastern Europeans, were hungry for meat and sugar for aside from a pig each family killed around December to furnish lard and meat for the upcoming year, very little meat was seen during the usual Eastern European year. The poultry the womenfolk raised had to be sold to take care of clothing and taxes. Bread for the year was provided by the shares received by the men for harvesting the landlord's crop. Sugar was a luxury seldom tasted by Eastern European peasants, and coffee was given to a woman only occasionally at childbirth or to a sick person about to die. And this hard diet usually resulted, within a year or two, in digestive troubles.

The second Saturday, I received my first pay from the steel mill. After squaring my accounts in the two Portsmouth boarding houses, I had $4.00 left. Not then knowing about the existence of Hungarian newspapers in the United States, but still wishing to know what went on in the world, I mailed $1.00 to Budapest and ordered the *Az Ujsag*, a liberal Hungarian daily for three months. The Sunday after my second payday was the end of the month, and it was time to pay for our board. This payment process was an unforgettable experience.

After the Sunday dinner, all of us boarders remained at the table, and the oldest man, Mrs. Petre's father, took two account books from his daughter.

One book was from the butcher, the other from the saloonkeeper. The total of the two were added, and each man was requested to hand over $5.00 to the old man. My turn for payment came later for I did not start at the beginning of the month. The amount so collected was not enough, and an additional $1.00 was called for from each boarder. The total amount after collection of the additional dollar was too much by about $4.00. A frantic search was made for change, and $.25 finally was returned to everybody. Then $.04 was returned again by every man to make up the required amount. But the operation did not end there.

When it was determined that each man's share amounted to $5.71, half of the amount was returned to Mr. Petre for the boarding house bosses paid only a half share wherever company board was maintained. Luckily, I also had to pay a little over half month's share. Thus, my contribution served to offset the new shortage, or else it would have taken several new rounds of this business to make up the shortage created by the return of the half share to Petre. As it was, the few cents left over when I paid my share were spent for beer. On top of the $5.71, each man paid $3.00 to Mrs. Petre for cooking, washing, and the shelter. The whole operation took about four hours and nothing would have induced me to tell them that I could do it on paper in four minutes for smart alecks were not welcome or popular in Hungarian boarding houses.

On my next payday, I went on a buying spree by purchasing a "harmonica" suit case, a pair of white pants, white canvas shoes, and overalls for home wear. It was a great mistake to dress up thus for the men wore clean overalls on Sundays and my whites were sharply disapproved. By that time, though, I felt that I was sufficiently established, and I did not care much whether they approved of me or not.

Soon a new and serious cleavage was created between my fellow boarders and myself by the arrival of the newspapers from Hungary. Our mail came in care of the steel company, the newspapers were handed to me by the foreman, and his interest was invited by a "foreigner" who received a daily newspaper from overseas. The newspaper was quite new in his experience, and he started to talk to me occasionally in German. The few conversations resulted in an invitation to a Sunday dinner in his home. I could not very well refuse, and I knew that my fellow boarders would not like this new friendship of mine. My very effort to work a little harder than the rest to prove myself was misunderstood from the start. They felt that I was trying to work myself into the good graces of the foreman and, if proof of this were needed, here it was.

Upon my return from the Sunday dinner, a few bitter remarks were made, especially by those who "felt good" toward Sunday evening from the consumption of a goodly amount of beer. And to make matters worse, next morning I was transferred to the night shift principally because it was discovered that I understood some German and could translate the orders from the night foreman to the others. But the night shift crew had to work twelve hours which meant that I made $.30 more than my fellow boarders, enough to cover the cost of the board. All of the men turned against me, and they did not fail to express their opinions about foremen's pets. They also knew that the night shift crew was just for emergency and that we worked only about half the time of the shift and were permitted catnaps between jobs. The only tough part of the night shift was the period at midnight when we had to get down to a deep pit under the rollers to load out the accumulated slag.[9] It was hot in the pit, the slag also was hot, and we had to re-sole our shoes every few days. But sole leather was cheap, $.10 per pair of shoes in the ten-cent store.

I took two egg sandwiches to work, and the old man of the company set the price of $.10 on them, which was almost half the cost of the regular monthly board, to get even with me. During lunch time, enough racket was made purposely to disturb me in my sleep. When I went to sleep in the yard under a willow tree, a few boarders always managed to have a load of conversation and wake me up.

Finally, I got good and mad and went to my foreman friend to request work form 1:00 P.M. to 5:00 P.M. on the day shift in addition to my regular night shift work. He was not surprised for foreigners were hogs for work, and he knew that we did not kill ourselves on the night shift. And to the chagrin of my fellow boarders, from then on, I worked seventeen hours a day and made $2.55 against their $1.50. This was an unforgivable sin by an upstart.

Meantime, slowly, all too slowly, I became acquainted with some American customs and institutions. There was a small park a few yards from our home, and I spent all my free time there. I watched the merry-go-round with its mechanical horses dipping up and down, but I was especially intrigued by the usual soda stand and the ice cream soda. Once, I stood around until someone ordered one and, quickly jumping on the stool next to him, I indicated by sign that I wished to have the same. The price was then $.05. I often repeated the performance for the soda tasted good. It was so much easier to buy popcorn for I just had to point to a bag and fork over my nickel.

I was happy and contented. I made good money, 12.75 Kronen a day. But by that time, I knew that I never would return to Hungary, and I thus started to count money in dollars, except when I compared my earnings to those of Géza. Géza had an excellent position, was much respected by the community and the family, and he worked for 200 Kronen a month against my over 300 Kronen a month earnings. I became acquainted with other Hungarians living in the other boarding houses and also with some German-speaking young Americans who told me much about America.

I always will remember a letter I wrote to Géza when he reminded me about the fact that I was a common laborer and told me that he could not see my enthusiasm about America. I told him about some aspects of the American style of life (as it was then), not failing to remind him of Hungary's system. I pointed out that in Hungary, even comparatively well-paid office workers were unable to keep both ends met for in "living up" to their position they were ever indebted to tailors, shoemakers, and others, and young men usually paid their accumulated indebtedness from the dowry they received upon marriage. I told him that as against that, Americans would be disgraced to buy anything on credit, except a home, a farm, or furniture if they were newlyweds.

How could I have dreamt that one day automobiles and hundreds of gadgets would be sold on installments in the United States, and that people would be high-pressured into opening charge accounts in various stores and would be enticed to spend tomorrow's earnings today? I often wonder whether we Americans are happier for this new style of life adopted shortly after the First World War. I especially was wondering during the years of the Depression when tens of thousands of honest American workers were forced to seek relief in bankruptcy.

Besides the ice cream soda and popcorn, I also explored grocery stores and was happy to discover the delicacies obtainable in cans. Once I pointed out to the grocer a can of cabbage, and I enjoyed the raw sauerkraut when he opened it for me. I also had him open a jarful of pickles and took home parts of my new discoveries. From then on, we had sauerkraut, pickles, and ketchup with the everlasting meat courses. But while the new fare was enjoyed by my colleagues, they were resentful of accepting same as my discoveries and were worried about the additional costs.

During my stay at home before noon, I became quite friendly with Mrs. Petre. I was, and still am, very fond of children, and reached her heart by playing with the kids. I was sorry for the hardworking girl with her third child on the way, and I often carried her pail of water from the well. She (as well as all other immigrant women) had to work hard, beyond anything known to the modern American woman. She had to bake bread for twenty-eight of us, including the family, and wash the dirty underwear and shirts for work and a shirt for each of us for home wear. She had to make up the beds and once a month had to wash the gingham that served as linen on the beds. Plus, she had to occasionally sweep the rooms and always take care of the children. Petre was a good boy and helped her, but, of course, he worked in the mill and could only do some chores after quitting time. The brunt was carried by the young woman. Once I asked Mrs. Petre to iron my Sunday shirt, offering of course to pay for the special work. She did not know what ironing was.

Once, Mrs. Petre told me about her secret ambition. Before she left for America, another woman from her village mailed home a photograph showing the other American women wearing hats. Hats in Hungary were worn only by ladies, and a peasant woman, of course, was not a lady. A Hungarian peasant woman wore a neckerchief or shawl on her head. Mrs. Petre had the burning ambition to acquire a hat, but she was afraid. Not so much afraid of her husband, who was a fine boy and treated her well, as of her father and the rest of the clan. With the exception of myself, all of the clan came from the same village, and they all were related to each other in some manner. I encouraged Mrs. Petre by pointing out that she was working more than anyone else and that she was making more money than her father or her husband.

I finally talked her into going to town while I took care of the children. Mrs. Petre returned with a shining face, and she was all dressed up in a large hat decorated by a long ostrich feather. The feather was at the front of the hat, sticking vertically, and bent back at the top. She tried on the hat for me, putting it on backward, and at each step the feather made a sort of bow, reminding me of the similarly decorated horses in the funeral corteges of the "Enterprise des Pomes Funabres" of Budapest, a ritzy undertaker. But she was so happy in her new possession that I did not have the heart to point out to her the backward error, and for all that I know, she always wore the hat backwards if she did wear same. At any rate, the hat did not make its appearance in public while I lived at the Petre home, though Mrs. Petre must have shown it to her husband in their privacy.

As everything else in life, the happy days of Portsmouth came to an end in about ten weeks. While the Petre family and some of the boarders warmed up to me and some even had me read the *Az Ujsag* aloud to them or write letters to their parents or wives, most of them became more antagonistic toward me than they were at the beginning of my stay amongst them. My utmost efforts to gain their friendship were in vain. They resented my earning more, my growing acquaintance with Americans, even my growing friendship with the Petres. The boarders did not try to conceal their dislike. I also missed my colleagues of the *SS Carpathia*, all of whom had left by this time for larger cities. Dezso Marosi wrote to me that he was in Cleveland where there were thousands of Hungarians and where whole streets were populated solely by Hungarians with dozens of stores owned by or conducted by our countrymen.

And just as I became restless and about to take a plunge, a Hungarian came to Portsmouth from Toledo, Ohio, to inform us that the pay of common laborers was $.17 there against the $.15 we received in Portsmouth. I decided to leave for Toledo and served notice on the foreman and Mrs. Petre. Both of them tried to keep me there. The foreman explained that the following year we would have a national election and steel plants might shut down all over the country. However, the Portsmouth Steel Company had orders for over three years. In my superiority, I smiled at the childish take and the crude story of the foreman. He spoke as if we did not have elections in Hungary when the government party was always re-affirmed in its power! For the life of me, I could not see what an election might have to do with the orderly flow of life and with productions in factories.

To the unconcealed joy of the boarders and the sincere regret of the Petres, I said good-bye on a Saturday night and withdrew my savings from the national bank of all immigrants—the undershirts of Mrs. Petre. I arrived in Toledo about 5:00 A.M. the next morning with over $30.00 in my pocket. My first trip took me to a lunchroom and, noting another guest eating strange food, I indicated by sign that I also wished to have a helping of corn flakes. Corn flakes were a new and pleasant addition in my store of knowledge.

About 7:00 A.M., I walked the approximately four miles to the Toledo Hungarian section—the man in Portsmouth had told me to ask for Genesee Street—and I tried to obtain board in the first two houses I went to. Perhaps because my harmonica suitcase did not look good, or perhaps because I failed to make a good impression in my white pants, I received unfriendly receptions at both places. I decided to proceed to Cleveland by trolley. The mere fact that trolleys were running such distances intrigued me, and this also may have had something to do with my quick decision. The trip was very nice since the trolley tracks were running on the Erie lakeshore.

CHAPTER 5

Cleveland, Ohio

★

I ARRIVED AT THE Public Square in Cleveland about 4:00 P.M., and my little green book told me how to inquire about Hungarians. I have no idea how I did pronounce the question, but the policeman on the square must have been used to Hungarian accents and pointed out the Hungarian Inn nearby. It was, what I now believe, a third-rate hotel, but to me it looked the utmost in luxury. I never was inside a hotel before that day in Cleveland, and I was accustomed only to the accommodations of Hungarian boarding houses. The hotel with its running water and white linen looked like Heaven to me. My respect for America was increased, but my joy somewhat dimmed when I learned that the room would cost me $1.00 a day.

The hotel owner was a decent and kind man. When I inquired about the possibilities of obtaining work and told him that I was looking for common labor, he advised me to look up the Hungarian section of Buckeye Road. He also advised that I should look for cheaper quarters. Perhaps his kindness and good advice were coupled with his business judgment that a common laborer in his dirty working clothes would not fit in with the rest of the hotel guests.

Anyhow, I spent a very pleasant evening at the hotel and got acquainted with a noted Hungarian actress, Laura Réthy, and with her husband, the Hungarian poet Gyula Rudnyánszky. Both are deceased long ago. Rudnyánszky advised me to seek employment with some Hungarian steamship agent and private banker as an office clerk, but he also told me that such jobs would pay hardly more than $6.00–$7.00 weekly. I decided against Rudnyánszky's advice. Nor did I take the advice of the hotel keeper who told me to seek lighter work such as washing dishes or being a bus-boy. I wanted to prove to myself and to my family that I was not afraid of real work and would not join the immigrant ex-army officers in menial tasks.

Next day, I moved to the Hotel Budapest on Buckeye Road where the room cost only $2.00 per week. But I considered that also too much for in Portsmouth, the whole board did not run over about $2.25 per week.

I did not immediately find my friend Marosi—and did not find him for over twenty-five years. But I did find the Hungarian daily newspaper, *Szabadság*, which was and is still being published in Cleveland and to which—incidentally—I am now contributing a column each week. With the *Szabadság* in my hands, I knew that I would thereafter not be a lost soul in an utterly strange country.

The owner of the Hotel Budapest advised me to try for a job at the Kuntz factory which was owned by a Hungarian and readily employed Hungarians. But when I learned from others that Kuntz paid only $.13 per hour for beginners, I decided against working for one of my countrymen.

From other Hotel Budapest tenants, I heard about what was called the Old Factory, a steel mill and foundry which was located nearby and where the wage was $.17 per hour. The tenants also told me that it was a notorious plant; the job was man-killing. But, if one worked fifteen hours a day, the pay was $.20 per hour, all of $3.00 a day.

I never have made easy money throughout my stay of over forty years in the United States, but I never worked as hard in my life as I did in the Old Factory. And my sentiment was shared by almost everyone working there for whomever I talked to had started to work there on the same day, or the previous Monday, and expected to quit his job the same evening, or the next payday. I know this sounds strange today when labor turnover is a great problem and is to be avoided by intelligent management. But immigrants were unloaded daily by the thousands in those days, and labor was plentiful for all purposes.

My first job at the Old Factory took me to the casting room. Casting is done by molders who build the necessary forms from sand on a wooden belt or plat. When forms are ready, the molten iron, called hitz, is poured into them. My job was to pick up the heavy forms then dump them to separate the casting from the sand. The sand was still hot and burnt my hands when I picked up the heavy forms. When the foreman gave me a pair of gloves, the burns were even worse. Some sand always managed to get into the glove and burn blisters on my hands. Therefore, I chose to do away with the gloves.

A few hours later, I was given the easier work of loading iron dust from under the grinders into wagons to be hauled away from the shop. The size of the shovel scared me for it was much larger than the ones used in the Thacker Mines, and, of course, the iron dust was much heavier than the coal. I do not know how many drops of tears were mixed in the dust. I was blaming myself as a darned fool for leaving Portsmouth for my hindsight was always better than my foresight. But somehow I managed to last fifteen hours, and for the first time in my life, made $3.00 in one day. I dragged myself somehow to the Hotel Budapest, four blocks away, and after some rest and supper, I went to Holton Street to shop for board. It was easy to secure board at the first house, Mrs. Frater's, for no letters of recommendations were needed for a man who stated that he worked in the Old Factory.

In Mrs. Frater's house, I got acquainted with another system of providing board for immigrants. The owner provided everything but the meat for the boarders. However, each man had to have his own book from the butcher and had to tell his daily meat wish to Mrs. Frater each morning. Some boarders wanted eggs and some wanted pork for breakfast, some wanted ham and some wanted veal for their lunch buckets, and the poor woman had to buy and prepare ten to twelve different things for so many boarders. Bread, soup, vegetables, coffee, etc., were provided for everyone, and quite a few men took advantage of the system by buying very little meat and gorging themselves on bread, soup, and coffee. Such men had to move often. Even with the new meat system and with the better balanced diet at Mrs. Frater's, the board did not amount to more than about $12.00 monthly for immigrants did not buy from Park and Tilford[1] class stores. Soup meat in those days cost about $.04 a pound, and we paid about $.09–$.10 for a pound of ham. Boarders often moved from one boarding house to the other to save $1.00 or even $.50 per month. Each immigrant came for a temporary stay, and each penny saved took him so much nearer to the day of his return to his country and to his loved ones.

I was unable to keep up the pace of working fifteen hours every day, so I worked alternately twelve and fifteen hours a day. I always played with the idea of quitting and seeking another job in another factory. It was hard to get up in the mornings and especially hard on Monday mornings. When doing heavy manual labor, my muscles slowly adjusted to the work, and after a period of rest, the re-adjustment was even more painful.

When I reached the third week of work at the Old Factory, I was considered a veteran in the plant. Therefore, I was entrusted with the toughest and most dangerous work, that of carrying hitz from the furnace to the molders. A large bull ladle that held twelve to fifteen gallons of hitz fit into a round hole in the middle of a twelve-foot long steel frame. Both ends of the frame were forked, and a man stepped into each fork, like a horse into a buggy, and thus two men jointly carried the heavy load. We did not have far to carry the load, only about 150 feet, but the distance was more than enough. The load was not only heavy; it also was burning hot. I was always afraid that either I or the man at the other end of the fork would stumble and fall, thus causing certain burning and perhaps death for both fork carriers. I wore three shirts and two overalls to withstand the heat and, of course, we sweated profusely. After each trip to the molders, I went to the yard, took the water hose into my mouth, and drank as much as I could. I still have the creeps when I remember those days.

On the afternoon of the fifth day at this job, I collapsed and was taken to a hospital. Four days later, I learned from Mrs. Frater, who visited me each day, that I was stricken with pneumonia. It was far from pleasant to lie in the hospital for over two weeks when I could not speak English to anyone and did not understand people who tried to speak to me. When I was discharged, I was too weak to look for work. Since I previously had replenished my scant wardrobe, I did not have over $30.00 cash. And winter was coming.

I usually sat at the Erie lakefront on nice fall days and watched the large boats leaving. I knew that they were not sea-going ships, but they reminded me of the sea, of the *SS Carpathia*, and of home. I felt forlorn and, for the first time, homesick. That fall, I passed my nineteenth birthday and, at that age, it is tough to be alone, sick, unable to work, and amongst strangers in a strange country. I cannot blame myself even today that occasionally I cried, but I remembered the old adage, and I always managed to cry alone where no one saw me. I also decided that come what may, I would have to learn some English.

A Mr. Brown was advertising his English course at $1.50 a week, and I enrolled. He was not a lucrative businessman for there were only five of us taking his course. I had learned very little English in America for when one lives amongst his own countrymen, one follows the path of least resistance, just like the water in its flow, and what one learns to speak at first arrival in America—as I have said before—is anything but English. And the English language is very hard. But, even if I learned only a little in Mr. Brown's school, I was still lucky to have enrolled. One of my fellow students secured me a job fit to my weakened physical condition. Within a few days, I was washing windows in the Hotel Hollenden. It was easy work, indeed.

The Hollenden forelady assigned five rooms to me, and besides washing the windows in them, I had to water the flowers on the porch daily. I never will forget the kindness and patience of the forelady. When she told me to take a pail, I did not know what she was telling me to do. She took my hand, led me to the pail, pointed to it, and repeated at least ten times, "pail," "pail." Then she had me repeat "pail" a few times and to this day, I know what a pail is. Then she took me to the first hotel room, taught me how to clean the window, and pointed out the five rooms that I was expected to attend to.

In my eagerness to make good, I hurried with the work and upon finishing the five rooms, I reported to the forelady. I had then the surprise of my life.

When the forelady discovered that I understood some German, she combined her good English and her poor German, and she gave me a call down for trying to hog up more work than I was expected to do. "Leben und leben lassen" ("Live and let live") was the motto of her lecture, and she ordered me to start over again. She explained that all I was expected to do per day was water the flowers and clean the windows of the five rooms.

A call to lunch put a stop to the forelady's lecture, and I got acquainted with a bunch of my fellow workers at the lunch table. When our shift ended at 4:00 P.M., we were fed once more. In addition to the three meals and the numerous snacks that the hotel Hollenden provided to us, we received $18.00 monthly. Of course, the food was strange American food, but through the kindness of the cook, or the man in charge of the pantries, some coffee, milk, or cakes were always available. I took full advantage of this opportunity and ate as much as only a nineteen-year-old boy recovering from an illness can gorge. The trouble was that evenings were long and at 8:00 P.M., I was hungry again. Since I had to pay $1.25 for my room and $1.50 for my tutoring to Mr. Brown, in addition to occasional tram fare, there was very little left of my wages.

I learned to skip paying tram fares. In 1907, neither the public nor the conductors were insulted by the tram companies' having a box to receive the car fare. Conductors went over the tram car, collected the nickels, and rang down as much as they wanted or had time to ring. If conductors were "too busy" collecting and returned after their shift with less than the money expected of them, they were simply fired, and the superintendent thanked them for at least returning the tram car. Cars were open and had a car-length board for a step. At 5:00 P.M., people were hanging on the board by the dozen. One simply took position at one end of the board and by the time the collecting conductor fought his way there, one got off the car before he was reached by the conductor. Some days, I rode three or even four cars to reach home, but the nickel was saved.

Once a week, each man working at the Hotel Hollenden had to return by 7:00 P.M. to stand watch in the laundry. To this day, I do not know what we watched for, but I have a vague idea the watching was to prevent fires. Those days were the brightest days for me because girls worked by the dozen in the laundry. I was nineteen years old, and the girls in the laundry were the first girls whom I'd been in a position to approach since I left Hungary.

Americans were not aware of the problem of no immigrant girlfriends for the young immigrant men between the ages of eighteen and thirty years old who yearly came by the millions to America. These young men found very few girls of their own nationality in America because, for example, there were any number of large Hungarian colonies without one single girl. Families keeping Hungarian boarders were young families and had no daughters old enough to be someone's girlfriend. There was no girl, for instance, at Portsmouth. The comparatively few immigrant girls in America usually obtained employment as domestics in New York, or at the Johnson & Johnson medical supply manufacturing firm in New Brunswick, New Jersey, or in the textile plants at Passaic, New Jersey. The immigrant men were scattered in coal mines and in the steel towns of Pennsylvania. Since immigrant men were unable to speak English, they were afraid to approach American girls for either companionship or for more ulterior purposes. Some of the men boarders besieged the wives of the boarding houses, and tragedies often resulted as a consequence of the unbalanced situation. The great majority of the immigrant men suffered in silence, but suffer they did.

Lots of single men could not wait for their return to Hungary, but went instead to Passaic or New York and got married. This situation was not remedied until after the First World War. By that time, many single men went home to marry, and many others were Americanized to the extent of courting and marrying daughters of other Hungarian immigrants.

All the girls at the Hollenden were immigrants or daughters of immigrants, and they spoke English. I learned a lot more English from them than from Mr. Brown. One may imagine that our conversations were on different subjects than work in factories and such. But, pleasant as it is to think of those hours, it would do me no good at my present age to dwell on the subject too long.

One month later, I felt strong enough to look for a better-paying job again, and I needed money badly. With the decision that now I could get even with life and stay in the Hollenden as a paying guest, a decision I carried out at numerous times, I left my job and obtained employment in a small foundry.

I decided to learn molding for molders were the best-paid men, and I agreed to work at $1.25 a day while learning. I was to make molds for stove parts, and this was very light work in the trade for the parts were small and light. To mold stove parts, one has a small board on which to build the pattern of wet sand and leaves a hole on top to pour in the molted metal, the hitz.

Prior to the close of the working day and when the mold patterns were ready, I was to take around the hitz to the other workers as I did in the Old Factory. Using a long ladle, I was to pour the hitz into the molds. A few minutes later, other laborers—the molder is not a laborer, but a skilled man—would throw the forms down to separate the casting from the sand, just as I did on my first day in the Old Factory.

I had only three good castings out of forty-two molds on my first day at the foundry. This was not a very promising start. During the following weeks, my work improved slowly but not as fast as was expected. Workers starting about the same time as I were building about fifteen good patterns a day, out of which they received about one hundred good castings. However, my best was about one hundred and twenty patterns with not more than about thirty good castings. At the end of four weeks, I was notified that my student days were over. Instead of receiving $1.25 a day, I was to do piece work and be paid $.04 for each good casting. I was not elated about the advancement, and I was not happy, anyhow, for my fellow workers were gloating about my unsuccessful efforts and did not conceal their contempt at my poor achievements.

But I stuck to the foundry job for another week. One rainy day, a drop of water reached my ladle from the leaking roof. Water does a strange thing to melted iron. A thousand sparks resulted from that one drop of water that hit my ladle, and one of the sparks struck my knee, burned through my pants, and left a hole into which I could place a good-sized hazelnut. Compensations were unknown in those days, and when I was ready to seek other, safer work about a week later, I had less than $5.00.

Winter was close at hand, and I needed some winter clothes and underwear. Once more, I felt alone and forsaken and somewhat scared of life again. Luckily, I did not know at the time that the "panic" of 1907 was on and that manufacturing plants were closing all over America. Had I been aware of the events to come, I would have written home for passage money to Hungary. This money was often offered in letters from home but never considered by me.

I told Mrs. Frater that while I was looking for work, it would be more convenient for me to eat out. I would again be a roomer instead of a boarder. Usually I went to the end of the streetcar line to get a piece of bread with an occasional piece of cheese or bacon. Sometimes, I took advantage of the free lunches, but the saloons in the Hungarian districts were offering very poor selections, hardly worth the nickel. I had to pay for the beer I did not drink in the Hungarian saloons, and going downtown to better saloons meant carfare. I then realized that poor men always have to pay more for everything than the rich.

Day after day, week after week, I went around from plant to plant to beg for work but had no luck. Two weeks later, after my vain search for work and after paying my weekly room rent, I had $.40 left. When I returned home about noon, discouraged and tired, Mrs. Frater took me to task. She knew that I was not eating regularly and that I was losing weight. Without any preliminaries, she called me to the table and set a dish of goulash[2] in front of me. She scolded me and said that I was a silly boy and that no one needed to go hungry in America, even if one were out of work. She told me that boarding bosses were not afraid to carry their men when work was slack if the men were honest and willing to work. Mrs. Frater also said that I again would be on full board with Mr. Frater's paying my share if, at the end of the month, I were still out of work.

I was happy to wolf up the food that Mrs. Frater gave me—Were you ever very hungry at nineteen?—but far happier because of the trust placed in me. Finally, I had gained the confidence of the people with whom I associated, and I cried from sheer happiness.

I felt as I feel now, that those wonderful women, the hard-working wives of the immigrants, were never appreciated properly. They came to America in their youth. Upon arrival, they usually were confronted with a boarding house previously purchased by their husbands from another immigrant family who had returned home or moved elsewhere. And the day after the women's arrival, they were cooking, washing, and baking bread for twelve to twenty working men. They did this for ten to twenty years, in addition to giving birth to a child each year.

The women of the boarding houses had to get up at 5:00 A.M. to prepare breakfast and to bucket the lunches. They had to clean up the place, attend to the children, rush to the store where, unable to speak English, they had to point out the goods that they needed. Then, they had to return home in a hurry to carry innumerable pails of water from the well in order to wash dishes and clothes. The women had to have plenty of hot water for the returning men, bake the wonderful bread, each loaf the size of a cartwheel, serve supper, wash the dishes, feed the children, and collapse on the bed about 9:00 P.M.

And now, when those elderly and old women of the boarding houses are going around with thick ankles and their mongrel English, a few of them with only a shawl on their heads, they often are ridiculed and looked at with contempt. Without the heroic labor of these women, the millions of American immigrants would have been unable to stand the exile from their own countries, and those millions were sorely needed by the fast-developing American industries. When I much later became acquainted with some pioneer women of our West, I compared them to the women of the boarding houses and did not find any difference between the accomplishments of the two groups. And when I was privileged to serve my America overseas during the unpleasantness of World War II and saw the many officers and enlisted men with foreign names, I was again thinking of the many six-star wives of the Hunkies,³ Hunyaks,⁴ and Dagoes.⁵ I did not feel that the ladies of the Daughters of the American Revolution⁶ had anything on the immigrant and pioneer women.

I looked for work for more than a month, hoping against hope that I would land something, but the work "panic" was repeated over and over again. In America's early days, a spade was called a spade, and the elegant expression of Depression was not in vogue then.

Once, a plant was advertising for mechanics, and I told the foreman that I was one, trained in Hungary. European mechanics were better trained than American mechanics for European mechanics had to serve four years of apprenticeship. The foreman asked for my mechanic's papers. I told him indignantly that we were in America where one is not supposed to carry proofs, and he took me on. My job was to handle a power-drill, and everything went quite well for two days. I was hopeful that eventually I would learn the trade and, meanwhile, earn $2.25 a day. The third day, the drill stuck, I was standing there helpless, and, of course, I was fired.

Next, I answered a Hungarian *Szabadság* ad requesting a painter. The boss was a Hungarian, and he discovered within two hours that I never had held a paint brush in my hands. He was rough, unwilling to pay me for the two hours, but he finally came across with $.50 after much haggling. Since then, I was ever afraid to misrepresent my talents.

I was more than $15.00 in the hole when once more good fortune—or what I then believed to be good fortune—smiled on me. An insurance agent came to Mrs. Frater's home, and I recognized him as Mr. Ulmer who came from Pásztó. He was a rich man once, the owner of the only two-story building in Pásztó besides the Count's castle, the owner and operator of a large stone quarry, and a contractor. But, his businesses failed about two years before I set sail to America. There was some scandal about forged notes, etc., but in my joy of meeting someone from the hometown, I did not mind his past. Mr. Ulmer was likewise glad to meet me and invited me to live with his family. He was very poor, eked out a bare living by selling insurance to immigrants, and was supported by the occasional work of his son, Nicholas, one of my former schoolmates.

Mrs. Frater did not mind my moving for they were carrying several boarders on the cuff, and she knew that I would pay her as soon as I were able. I moved in with the Ulmers.

CHAPTER 6

Again In Coal Mine

★

MR. ULMER, NICHOLAS, and I went to look for work for four days. On the fifth day, there was an ad in the *Szabadság* for coal miners. Men had to apply in person, ready for work, in a Hungarian restaurant in Pittsburgh, Pennsylvania. I talked Nicholas into joining me in a trip to Pittsburgh. Knowing that his parents would object, we decided that we would not say anything at home but would take a freight for Pittsburgh and write for our clothes from the mine.

Next morning, we did go out to the railroad yard. I did jump on a train, but Nicholas, who did not like the prospect of mining, anyhow, refused to follow me, and I went alone. I had $.10 in my pocket and carried one extra shirt.

The train took me as far as Youngstown, Ohio. From there, I walked and slept in empty box cars on sidings for two nights. I had nothing to eat on the third day. Finally, I reached the bank of the Monongahela River opposite Pittsburgh. I was so hungry that I approached a gentleman to ask for a nickel for bread. He gave me $.02. When I left him, a young Negro who had overheard my request followed me and gave me $.05. Ever since then, I have had a soft spot in my heart for Negroes. Later, I realized that I would have been a friend of Negroes, anyhow. I carried two heavy inheritances, the ghetto and the poverty of the peasants of Mátraverebély in my heart, and I could not help but sympathize with every underprivileged man.

I was an immigrant, somewhat of an underdog, and therefore less than 100 percent American myself. I felt keenly the unjust and treasonable activities of the Ku Klux Klan and the similarly-minded people. While we immigrants may have been advised to return home because we came to America on our own volition, the Negroes did not fight very hard to crush America's gates when they were forcibly picked up by our Yankee forefathers.

At this point, I should be permitted to say something about we uninvited immigrants in order to clarify the immigrant issue to those knowing little about the history of our country. Immigrants were invited to these shores from the beginning of the American colonies right up to the end of the First World War. For decades, land companies flooded Europe with their invitations for settlers, and this included the land company organized under the leadership of George Washington. Following the Civil War, when our country was industrialized first directly by American industrial concerns and later indirectly through steamship companies, a vast propaganda for immigrants was carried on in the Eastern European countries. In fact, the propaganda was so intense that in some countries, representations against the propaganda were made to the American Legation.[1]

When Mr. Morgan's Steel Trust[2] imported thousands of Poles during the bloody strike of Homestead,[3] the example was followed by other industries until Congress had to pass laws prohibiting the landing of immigrants who arrived in America as contracted laborers. Even in my early days, it was a common practice for foremen to approach good workers and offer advance pay for steamship fare if one or more of the workers' relatives could be induced to come to America from the Old Country. We immigrants suffered a lot from unjust discriminations, such as our being told in days of Depression to go back from where we came. It is high time that I speak up and enlighten America's native citizens that we immigrants were indeed invited to America's shores.

It would be tempting to discuss the merits and demerits of immigrants from Eastern Europe, especially after the stigma of the quota system which gave the unjust preference to so called "Nordics" and thus established racial theories in America much before Hitler. But this discussion would require a book-length symposium.

Back in Youngstown, I spent $.05 for a bread, but I was still hungry. Therefore, the other $.02 went for rolls for which I was very sorry the following morning after another night spent in a box car.

At the bridge leading to Pittsburgh, I was confronted with a demand for $.01 toll, and I did not have $.01. I had to walk several miles to a railroad bridge, and I crossed the bridge in a frenzy, lest a train should come and compel me to jump into the Monongahela River. Once on the other side of the river, I had to walk back toward the city. It was about 1:00 P.M. when I reached my destination, the Hungarian restaurant.

I sat down at a table, and the restaurant owner asked me whether I wanted soup or not. I was very hungry, of course, and did not have strength to tell him that I had no money. When I finished the soup and several slices of bread, I told the owner that I came to the restaurant about the mining job, and I did not have money for food. I thought that he would either kick me out for accepting the plate of soup or bring on the next courses. He did neither. He said that he was waiting for another man and in about two hours, he would take us to the coal mine.

The other man came and at 4:00 P.M., we were in Iselin, Pennsylvania, a small mining town. We were assigned to a Hungarian boarding house. We also were told that we would have to work on the night shift starting the following night. My new buddy was a middle-aged man named Lajtos who was an experienced coal miner. We bunked together during the close to six weeks that I was working in the Iselin Mine.

When Lajtos and I reached our working place, and before we started to work, I told him that I knew very little about mining, and I offered two-thirds of our loaded cars to him. My buddy was not surprised by my confession and said only that he would see what a fair division would be at the end of the first shift.

Coal mining in Iselin was different from mining in the Thacker Mines. At Thacker, the coal was shot down during the night, and the night miners' only job was to load the cars. In the Iselin Mine, we had to drill holes and shoot down the coal before loading, and this process required two men working together.

I learned at Iselin the reason for calling the auger the "cramp." The six-foot auger is forked at the end and adjusting the fork to one's breast and pushing the fork with one's might results in one's turning the fork around by hand. The auger has to be taken out of the coal wall several times to file the dull end before the six-foot hole is ready for the cartridge. If a thin sulfur vein were reached, one's buddy had to help with pushing the auger by adjusting his behind to the other fellow's behind and pushing him. When one had drilled two or three holes, one had cramps, indeed. I was very thankful to have an experienced miner for a buddy for I would have been unable to even start the mining process by myself.

When two or three drilled holes were ready, and an experienced and skilled miner was required to decide how many holes were needed, the miners wound a large piece of paper around the pick handle and filled the form with black powder. Before the cartridges were fired, the miners had to seek safety behind the bank car or run out of the working place that is called the "room." When the coal was shot down, we went to work loading. I tried to do my best to keep up with Lajtos. After loading thirteen bank cars, we still faced the hardest part of the shift. There was a vein of about fourteen inches of loose stone between the coal and the solid sandstone roof, and after cleaning the coal, we had to shoot down and pile up the stone along the ribs.[4] And although the holes were much shorter in the soft stone, the drilling was much harder.

We finished about 4:00 A.M. and made about $3.50 for each of us. Lajtos told me that he would accept only seven cars of the thirteen for I was a good and willing worker. Of course, I was happy to make money once more but far happier by Lajtos' statement. I began to feel that eventually I would be a dyed-in-the-wool coal miner and would cease to be an outsider. I was forever an outsider in the past years of my life, and unfortunately, I felt like one for over thirty-seven years until I entered the Office of Strategic Services[5] (OSS) and traveled overseas during World War II.

I was an outsider when I was a kid in Mátraverebély because I was the only Jew. I was an outsider even in the Jewish Normal School for I did not feel that I belonged there, and I was an outsider in the offices where I worked in Hungary because I was not a graduate of the Academy of Commerce. Of course, I now was an outsider amongst the Hungarian workers not only because I was born a Jew, but also because I was not from the working class.

I worked very hard in Iselin, but I never spent happier days during my first thirty-seven years of stay in the United States than the days in Iselin. Everything went well for about four weeks, but then the Depression panic also reached Iselin. Work became slack; we had only about two or three days of work per week. The mine was not a closed shop, but there was a union there, and the union organization became determined to force men to join or to leave the mine.

It would have cost me about $30.00 to join the union, board was $16.00 a month, and with the prospect of making about $6.00 a week, my finances did not look very promising. By that time, I felt that I was equal to work offered to us the day of our arrival in New York. The offered work was in the McAdoo Tunnel, under the Hudson River, in compressed air, at $2.00 for an eight-hour shift. My *SS Carpathia* colleagues and I had heard about the bleeding noses and ears that resulted from McAdoo Tunnel work, and we initially were afraid to accept the job. Since I had gained the respect of Lajtos as a hard worker and was now really used to work, I was no longer afraid of the McAdoo. Therefore, I left Iselin with $40.00 in my pocket.

I was somewhat disturbed that my letters to the Ulmers were unanswered. But when I left Iselin, I was glad that I did not have to carry my clothing because, whenever possible, I was riding the rods. Occasionally, I had to walk. At night, I always slept in boxcars and at Newark, I took the train.

When I arrived at the Rakoczi Hall in New York, I was greeted by old friends. Eger, Rahman, and Bakron from the *SS Carpathia* were glad to see me for misery likes company, and they all were broke. When I told them my plan about work in the tunnel, they laughed in my face and told me that one could not buy a job there, although pay was now reduced to $1.50 a day.

For the first time, I called on my Weiss cousin in Brooklyn to look for my mail for letters to me came to him from the Old County, and he forwarded the letters to me when requested. I was surprised to find a dilapidated frame house at my cousin's address. My cousin was somewhat embarrassed by my question about the manufacturing plant and told me that his business was to gather fat at restaurants and butcher shops for a soap factory. His equipment was a wagon and an old horse. The letterhead was to impress his business connections and his friends. He was a poor man, barely supporting his five children.

My cousin's failure to wait for me at the steamer was not mentioned. When he invited me to move in with his family, his wife jumped on him in Yiddish, a language with which I am not familiar, and I declined. In fact, I declined even the cup of coffee offered by Mrs. Weiss, and I left his house with several letters in my pocket.

One of the letters contained a very painful surprise. My oldest brother wrote to me that ever since I was fourteen years old, I had brought nothing but shame on my family and departing to America only put the crowning shame of all on them. I was dizzy and mad clear through, for I never wrote home anything but that I was working and satisfied. My family never knew anything about my sufferings.

But as I read further, I understood. Nicholas Ulmer, the filthy coward that he was, did not tell his parents about our plan to go to the coal mine. He was afraid. And for all the Ulmers knew, I simply skipped without paying for four days of board and even without saying good-bye. And old man Ulmer had written to my brother and had told my brother what an ungrateful wretch that I was. Ulmer said that he had taken me into his home when I was down and out, but I skipped and left all my belongings, even my photograph of my mother. Of course, Ulmer named the amount that I owed the Ulmers, about three times as much as was due. This amount had been promptly remitted by my brother.

Of course, a few days later, Ulmer knew that I did not intend to cheat him for he must have received my letters from Iselin. However, he did not have the decency of writing another letter to my brother. It took me years to convince my folks that I did not try to cheat Ulmer and that I lived honestly and never intentionally cheated anyone in my life. I was very grateful to my brother for also stating that my mother was spared the agony and was not told about the letter from Mr. Ulmer.

Much later, I learned that Ulmer and his wife and daughter returned to Hungary after Nicholas disappeared in the United States. Following World War I, Mrs. Ulmer wrote me from Hungary and asked me for a $2,000.00 donation for a dowry for her daughter. I did send the $2,000.00.

I deposited my $40.00 from Iselin with the Kollath Bank on Avenue C. Three days later, Mr. Kollath shot himself. For my $40.00, I received some education about private banks and private bankers. Scores of private bankers were advertising in the foreign-language newspapers and offering their services with sending money to the Old Country, safeguarding deposits, and selling imported articles. Competition was keen among the private bankers. Most of them conducted their business at a loss and made good the daily shortage from the in-coming money on the next day until the panic caught up with them. Many millions of dollars were lost by the immigrants with those foreign private bankers. Years later, the various states put the bankers under stringent supervisions. A few private bankers had the decency to kill themselves. More of them simply announced that they were unable to meet demands and shut their doors. Some failed with as much as a million dollars. Some left their "bank" penniless. Others kept enough money to start anew somewhere else.

Within a few days, I was stone broke in New York. I was broke to the extent of not having a roof over my head nor any clothing except what I wore, and I did not have $.10 to pay for shelter in a flophouse on the Bowery.[6] But I was not alone on that bitter winter of 1907-08. There were hundreds of thousands like me in New York alone, and hundreds of Hungarians in the various Hungarian saloons, such as the Rakoczi Hall, the Kossuth Hall, etc., where we were tolerated during the day around the stove.

In 1907, there was not a Civil Works Administration[7] or a Public Works Administration or any other organized or official method of relief. Everyone had to scratch for himself. Another twenty-five years had to pass and another panic had to hit America before Franklin Delano Roosevelt came to awaken our country to its social consciousness. For many months and years, people had survived on begging and stealing. It was during those horrible stealing and begging days that Judge Rosalsky of New York sentenced a man to an eight-year stretch for stealing a bread costing $.08.

Immigrants in America tried daily to secure a few pennies for something to eat during the horrible days. More immigrants were then coming in droves to America, and those of us already in America waited at Battery Place to guide the new immigrants or to carry their handbags for a few cents. But more often than not, we were chased away by a policeman who believed that we were trying to do the immigrants in. Some of us did.

Some of us, in fact, all of us, tried begging. There were some Hungarian businessmen willing to stake every Hungarian beggar to a nickel each week and one, a Mr. Malogh on Second Avenue, dished out dimes until he went broke and killed himself. A Mr. Wechsler, owner of a small dry goods store on Avenue A, was the kindest of all. He handed out some encouragements with his weekly nickel, and he accepted a very small parcel from me for safekeeping. It was the last little treasure I had, one of the poppy seed biscuits my mother cried her tears into on the day I left for America.

On red letter days, we Hungarian beggars were able to get enough money to sleep in a flophouse for $.10, but if one succeeded in having $.15, he paid for a bed in the Kossuth Hall on Sixth Street. In such rare cases, all of us *SS Carpathia* boys were taken care of. When the Kossuth Saloon closed at midnight and the owner went home, the paying guests came down to open the door and give all of us shelter until 6:00 A.M. the following day. When only a nickel was secured and none of us had $.15, I rode the subway all night or until I fell asleep and was kicked out. There were many of us taking advantage of the well-heated subway cars, and we were not encouraged to sleep on the trains. If I had $.03 for the night, I had to choose between sleeping and eating. Coffee was sold on the Bowery for $.02, and one was permitted to sit in the coffee shop until one fell asleep and was ordered out.

If I were not very hungry, I purchased a place in a moving van for $.03. The van was standing on Avenue B at the Fifth Street and housed about thirty of us nightly. But, we beggars also spent plenty of nights on the benches of Tompkins Square with plenty of newspapers under our pants and coats. We made numerous trips to the public toilets to warm up. New York nights in public parks are rather cool in January. On nights without shelter, we went down to the Bowery to the Salvation Army for the cup of hot black coffee (without sugar, of course) and the piece of bread that they started to hand out at midnight. Standing in line for about an hour or two, only to be told that there was no more bread when one had reached the window, was enough to bring tears into one's eyes.

Every morning, from 2:00 A.M. until 3:00 A.M., beggars went around the restaurants and nightclubs to forage in the garbage cans for pieces of bread or rolls. My favorite place was the Café Boulevard on Second Avenue. It was high class, and the garbage usually had several pieces of rolls in it. By 5:00 A.M., we were down at Park Row waiting for the German daily newspaper, *Staats-Zeitung*, a favorite advertising medium that usually requested foreign workers for porter or dishwasher jobs that included room and board for the services. We rushed to answer likely ads and often walked up to Eightieth or One Hundred and Twenty-Fifth Streets only to find that there were hundreds of us seeking employment. There were plenty of capitalists much nearer to the employer's door for they had the nickel to make the trip by the elevated trains.

When I was in the money, my favorite flophouse was the Hotel Alligator on the Bowery where shelter was obtainable for $.10. There were plenty of lice in the bunks, but we had no reason to mind that. Each of the guests took the rooms for the rooms also had showers with hot water day or night. Meantime, we were trying to get back on the *SS Carpathia* employment list and waited eagerly for her appearance, only to find again that the capitalists had all the advantages of the world for they could purchase the jobs that the foremen were selling for $5.00–$10.00.

I remember that once it occurred to me that for over five weeks, I did not taste anything but chunks of bread. I was always hungry, but not lazy, and I decided to go into business as an independent contractor. Starting at Sixth Street on the East Side, I went uptown and called in each saloon on every corner. I offered to clean up the place, including the men's room, for one dime. Some days I walked up as far as One Hundred and Twenty-Fifth

Street and, if I were lucky, I contracted as many as three jobs in a single day. But more often than not, the price paid for my work was only a nickel or food from the free lunch counter. And by the time that I had two or three jobs, I had been turned down, sometimes rudely, in hundreds of places. A nickel a day sustained me as far as food was concerned for a stale bread was sold at $.05. But the shelter was a daily problem.

One day, my buddies and I were electrified by an unbelievable sign in front of an employment office. The sign offered work in a coal mine with free transportation to the mine. Of course, we signed up as soon as the doors opened, and about forty of us were shipped the same day to Gleason, West Virginia, a small new mining camp on the border of Maryland. The forty included three of my colleagues from the *SS Carpathia*, the gentleman son of a Hungarian Lieutenant Governor, a black sheep of a very prominent Hungarian family, and a recent arrival from Hungary who also was broke. The recent arrival's name was Szuhay. Upon our arrival in Gleason, we were told that there were no Hungarian boarding families, and we were advised to establish bachelor housekeeping.

We were given a large house and separated ourselves into two groups. One group included all of the *SS Carpathia* fellows and Szuhay. Then, we started housekeeping. Furniture was given in the company store; also, such food as they had. Meat and bread were unavailable for three days, and we were existing on crackers, ham, sardines, and cookies. The superintendent was happy to receive such a bunch of Hungarian workers for Hunkies are known as hard workers. We were free to purchase anything in the store.

There were nine of us in our group, and no one knew the slightest things about cooking. Szuhay offered to remain home and cook provided that we share our earnings proportionally, and we accepted the offer. Upon our return from the mine the next evening, Szuhay admitted that the "soup" remained hot water and the potatoes a soft mass. We discharged him and decided that each day another man would remain home to clean the place and to try his hand at cooking. How could we not know that ready-made food was obtainable in cans?

We were not any luckier in the mines. The coal was only three feet high with several veins of dirt in it, and we were supposed to load clean coal. I was the only "expert" coal miner in the bunch. I had to drill and shoot the coal, and I was ever experimenting with the holes and the shots. To save my life, I could not load over two cars a day on my knees or on my belly.

Since everything looked black in the mine, I am sure that I loaded at least half as much dirt as coal. But the superintendent was patient and while my buddies did not do even as well as I did, our credit in the store remained good. We took full advantage of the credit by buying cookies, crackers, and other such delicacies with which we were familiar.

But the food remained an everlasting problem for each evening we were facing painful surprises with the "supper" prepared by the cook for the day. Thus, we kept on eating ham and bread which was ordered for us from the nearby town.

There was also the problem of washing, and to put an end to this story, one morning I found two of my colleagues AWOL. The following morning there were only three of us left: Eugene Bakron, John Fekete, and myself.

I did not wish to try again the days of agony of New York, so I talked my partners into seeking board with a Russian family. We moved there after returning the furniture to the company store and, if anything, the Russian home was worse than our own boarding house. The food was terrible, and the home was full of lice. Our accomplishments in the mine failed to improve while our indebtedness in the company store was increasing by leaps and bounds.

On a sunny winter Sunday, Eugene Bakron, John Fekete, and I grabbed a freight train heading toward Baltimore. We pulled in to Baltimore about 5:00 P.M. We were directed by a policeman to the Hungarian section and were lucky to find a Croat-Hungarian family. Black coffee and a piece of bread were given to each of us with the intelligence that the boarders were out of work and that work couldn't be had at all. We were permitted to sleep in a room completely bare of any furniture. Next morning, we were treated once more to black coffee and bread and were told politely that we were welcome to go.

It was rather early when we were turned out of the Croat-Hungarian boarding house, and as we walked the streets, a young Negro offering work approached us. The bargain was cinched down in a Negro lunch room with a plate of much-appreciated hot soup for each of us. We were led to a small boat going out for oysters. I was terribly cold at sea in my scant clothing and within minutes, I was seasick, much to the amusement of the men on board. As if the winter cold were not enough, I had to ice the oysters. My friends and I quit the oyster job the same afternoon after receiving $.60 each besides some food. We went to sleep in a flophouse.

Next morning, we called on the Austro-Hungarian Consul or Vice-Consul to request some assistance. I had heard previously that representatives of our country were authorized and possessed funds to assist Hungarian subjects in need. The Consul did not speak Hungarian, and when I told him in my broken German the object of our visit, he started lecturing me. He demanded our passports which we did not have. Finally, the Consul offered a quarter to each of us which was indignantly refused with the explanation that we were not bums but wished to travel to New York to seek employment there. With more lectures, he handed a dollar to each of us.

We were mad at the Consul and, to get even with him, we took the pair of shoes we saw in the anteroom and sold same for $.60 at a store selling second-hand goods. And to this day, I have no regret for stealing the shoes for stealing would not have occurred to any of us if the Consul had said one kind word in his two lectures. We took a freight train, and we expected to arrive in New York with some capital for the price of the shoes provided meals for the three of us.

The three of us climbed into the freight train's gondola car loaded with coal. We dug ourselves in against the winter wind with the pleasurable anticipation that we would arrive in New York with a dollar each, a rarity in those days in our circles in New York. Within about two hours, the train stopped at a large depot, and we judged by the many electric lights in the extended yard that we were in Philadelphia. No sooner did I stand up than I was invited by two men to climb down. Since both had guns, my friends and I had little choice in the matter. The men were railroad detectives, and we were arrested and escorted into the waiting room at Wilmington, Delaware. They went through our pockets and, of course, took the dollars from every one of us. They also tried to talk to us and question us.

I addressed the detectives in German and told them that my friends and I tried to go to New York to obtain some work. I refused to understand anything English, even when one of them stepped out and picked up two pieces of rocks to indicate that we were going to jail for thirty days and would break stones.

About half an hour later, a passenger train pulled in and the detective who took the money from us purchased three tickets for $.60 each and handed back $.40 to each of us. He led us out to the conductor, told the conductor something, and motioned for us to board the train. We did not know then that we were going to Philadelphia. This was one of the most, if not the most horrible hour or so that I ever spent in my life. I was fully convinced that the conductor would hand us over to the authorities to be locked up. Being an ex-convict is not a pleasant status in the United States, but is far worse in Europe where a man is requested to show his record at every step and where such things never can be eradicated. I was thinking of my mother, my father, my family—poor but honest people. If I had dared to leave my seat, I would have jumped out of the train in my effort to commit suicide. I blamed myself for being a smarty and leaving Hungary, for leaving Thacker, Iselin, and even for leaving Gleason. I cried very bitter tears the whole way to Philadelphia.

When my buddies and I reached Philadelphia, both the conductor and the brakeman went to another car, likely to let us go in peace, and go we did, running from the station much faster than I thought one could run. We ran until we reached a flophouse. I am still surprised that they took us in, covered as we were with coal dust. We did not mind paying $.25 for the lodging, a price far above the then-prevailing prices in similar establishments. After breakfast the next morning, we did not have a cent to our names.

My buddies and I knew about the so-called Jewish homes located in almost every city. The homes offered shelter and food for a couple of days for itinerant Jews. We went to the Jewish home in Philadelphia, and I had a strange experience there. I did not speak Yiddish, but Tekete, the Gentile boy at the home, spoke Yiddish. Following much questioning and with some doubts as to the fact that my buddies and I were really Jews, we were taken in for a single day.

Eugene Bakron, John Fekete, and I started the following morning toward New York. But after the lucky escape from the clutches of the law, we were afraid to try freight trains and thus walked. We slept in box cars but were educated by some of our bedfellows that knights of the road could have at least one good meal a day, and the trip was not so tough. Other hoboes showed us how to snitch milk and rolls early in the morning from porches in small towns, but we were admonished to never take both the milk and the rolls from the same porch. After all, the hoboes had ethics to which we also had to adhere.

In New York, we joined the boys who had left Gleason before us and took up life much as we left it before the excursion. Once, rumor started around the Kossuth Hall that plants were opening in Pittsburgh. At first, Bakron, Fekete, and I did not pay much attention to the rumors for such rumors started often, and they usually resulted in a cruel fleecing of some immigrants.

A man posing as a labor agent showed up at one or the other of the Hungarian saloons and stated that a number of men could have out-of-town work provided if they could pay their own travel fare, $5.00 or $6.00 a head. Some men who did have that much money volunteered for everyone was eager to work at anything. When the group was organized, the "agent" took them as far as Philadelphia and left them at the station. This game was played several times until close to fifty men were waiting hours at the station for the return of the agent who "just stepped out" until the other train arrived and a policeman got interested. Of course, the men were told that they were cheated. They had to return to their homes as best they could. But on the next occasion, one of such agents was arrested and his money was taken to provide return fare for the victims. The agent got four years in prison. This incident put a stop to that labor agent game.

But the rumor about the available jobs in Pittsburgh persisted, and I took the road once more for Pittsburgh. Of course, plants were not starting operation and the rumors proved to be baseless. I went to the Hungarian Consul's Office again for a few cents. While waiting for the Consul in the anteroom, I got acquainted with an Austrian boy of about my age who was there on the same errand. We had to wait a long time, and he told me his life story. He was to become an opera singer, but he was drinking and was booted out of school. His humiliated sisters provided steamship fare to America, the flotsam and jetsam of the world.

The Consul wanted to see our passports. Mike, the Austrian boy, had his and received $2.00 which was duly marked in his passport. I had left my passport with Mr. Ulmer; therefore, I was kicked out without getting anything.

My new friend, Mike, was a decent boy. He assured me that he would share his two bucks with me if we stuck together and looked for work in other cities. We started for Columbus, then for Cincinnati, and finally wound up as professional hoboes.

I never had a better friend and a more loyal buddy than Two Bits Mike. We went inside the public library in Cincinnati to rest and warm ourselves, and as we, of necessity, spoke German to each other, an old gentleman named Colonel Holden spoke to us and showed much interest in our stories. Later, Colonel Holden invited us to a lunchroom, treated us to coffee and rolls, and explained to us that he was a hobo. He was to leave for the warmer climate—it was early spring of 1908 and still rather cold in the Middle West—of New Orleans, and he invited us to join him.

CHAPTER 7

Riding the Rods

★

Two bits Mike and I were lonesome. We knew that without a skilled trade, we could not get work soon, so we took Colonel Holden up on the New Orleans offer. Colonel Holden, a gentleman if there ever was one, was a legendary figure in the hobo jungles in those days.

Colonel Holden was about sixty years old, a highly-educated man who always was reading German and French books in the libraries. He was a graduate of the University of Jena. I never learned what caused him to leave normal society, although there were rumors that he was disappointed in his family life. When Colonel Holden learned that I was a Hungarian, he took me back to the library and dug up some books to show me that Hungarians fought in the American Revolution and in the Civil War. He got me interested right then and there in the early history of our country, America, the study of which became my hobby in later years.

That night, Colonel Holden, Two Bits Mike, and I slept in the hobo jungle of Covington, Kentucky, on the opposite side of the Ohio River. Also on that night, I became acquainted with the institution of the hobo jungle, an institution that was soon gone with the winds and the automobiles. The hobo jungle was always located about two miles from any city or town, preferably under some trees or bushes. There were a few hoboes in almost every jungle in those panicky days, and the earliest arrivals took care of the food—always hot food, usually mulligan,[1] and always good. I never knew that so many men could cook rather well. Ingredients were secured either by purchase, or if necessary, by begging. Stealing was only tried when everyone was plumb broke. The hoboes were not bums or thieves.

The hoboes were men who got tired of organized society and normal life, men who wanted to forget their earlier environments, friends, and folks, if they did have folks, and who wanted to escape the restrictions and injunctions of society. Some carried heavy burdens, some knew penitentiaries and jails, but no man was ever asked questions about his past or person. Anyone was free to appear at the hobo fire of the jungle, free to partake of the food and drinks, if we did have drinks, which was the case most of the time. No questions were asked as to who the newcomer was, from whence he came, or to where he was headed.

The hoboes did not have roads in the country; perforce, we had to stick to the railroads whether riding the rods or walking. Automobiles and hitch-hiking did away with the proper hobo life and the hobo jungles. The jungles were one of the romantic aspects of the American way of life. Nowadays, one or another adventurer proclaims himself King of the hoboes and obtains write-ups in the newspapers and magazines. In my day, Colonel Holden was the undisputed doyen[2] of all the hoboes and the jungles, but he never called himself King. And still, Colonel Holden's word was the law wherever he happened to be.

If one of the hoboes needed assistance or if one of us got crippled jumping off the train, the whole hobo fraternity was mobilized by the Colonel to work and to hand over their earnings until there was enough to buy the cripple a small newspaper stand or cigar stand. We took care of the sick and the dead in similar manner, and we provided train fare and a small sum for many a youngster whom Colonel Holden talked into returning home to his parents. A new world was opened to me with the "door" of the first hobo jungle, a world I am still homesick after. It was a world of real camaraderie of unselfish men. There was no greed and no jealousy for no one had anything not shared willingly with others. Those having a few cents were willing to provide food and drinks. When work was obtainable and necessary, all of us worked.

In New Orleans, there were about thirty of us living in close fraternity with a bunch of fishermen. We were allowed to sleep on their small boats in exchange for assisting them and, if we wanted, all the fish was provided for our wonderful Louisiana fish stew. When a sea-going ship loaded with bananas or coffee arrived, the hoboes went to work making all of $.40 a day, far more than our daily need was. I soon had a new pair of blue denim pants and a jacket. We youngsters always looked clean.

Mike was called Two Bits because he was a gambler and ever invited folks to throw dice for two bits. I have to admit that his dice were loaded. When we were broke, we always depended on Two Bits's dice to make a few cents but never from other hoboes. There were no incentives to win money from our buddies for if anyone did have some money, the money was at the disposal of every other man in the jungle.

Two Bits was my closest friend, and we gathered unto us in New Orleans a small Mexican kid who was called Spotty because he was marked as a result of smallpox in his childhood. Three Finger Joe, a Canadian youngster, was the fourth addition. The four of us had a glorious time because Spotty was always playing nice music on his banjo, and Mike was a good singer. We enjoyed the fatherly attention of our older buddies because of our youth.

Our hobo gang did not stay put in New Orleans but made sorties, or trips, because we were ever wishing to see the other side of the mountain. We had a wonderful time in Savannah, Georgia, when the cotton was loaded. The town was chuckful of money, and Two Bits worked overtime with his dice amongst the workers—white and colored—of the docks. Sometimes the source of his luck was suspected, and we had to run. But we seldom got beat up.

While in Savannah, the Colonel was stricken with pneumonia and was taken to the free ward of the hospital. We did not like the idea of his being a charity case, but we did not have enough money notwithstanding Two Bits's operations. Thus, the four of us went around from saloon to saloon with Spotty playing his banjo, Two Bits and Three Finger singing, and my going around with my hat in my hand. We made over $7.00 the first evening but were arrested the next night and taken to the police station. We told the captain frankly that we were trying to make money in order to take care of our friend and somehow our story was published the next morning, with our pictures, in the local daily newspaper. Some ladies were deeply touched by our devotion to our old friend, and the Colonel was transferred to a private room at the expense of some local ladies' aid society. Two weeks later, he joined us in New Orleans.

Some hoboes, including the Colonel, were much interested in the Kentucky Derby. Nothing daunted about ten of us, and we took the train and went to Louisville to visit the races. While there, Two Bits was ordered to make enough money for admission tickets and a few bucks to play with for the Colonel. Victims were plentiful amongst the stable boys and other hangers-on. We forced the Colonel to buy himself a new second-hand suit and do the race in style.

In later years, I had my ups and downs. I was a coal baron once, a publisher at other times, but I never could afford to attend the Kentucky Derby except when I was a hobo. Is it any wonder that I sometimes long for the good old, carefree days?

Once, several members of my hobo clan decided to make a short visit to Florida, and we took the next freight. About ten miles out of Tallahassee, the most forsaken part of the United States, we were arrested. And we were arrested by no friendly railroad detectives like those in Wilmington, Delaware, but by a tough Southern deputy. We were sentenced to thirty days by the equally rough justice of the peace. The same day, we were leased out by the brother of the judge to work at a sawmill. And we had to work harder than we did at the unloading of ships. We were housed in a barn by the lessee and fed very uninviting and poor food and corn bread. The fourth day of our captivity, a liberated comrade who had served his sentence took out a letter addressed to the Colonel in New Orleans. The following day, the Colonel was with us. He demanded that we be freed and paid for the work and put up a fight that was something new in the judge's experience. The Colonel was well-known throughout the South and before leaving New Orleans, he had secured a letter from some authorities. Shaking the paper under the nose of the judge, the Colonel threatened the judge with dire consequences unless we be freed at once and cited Blackstone[3] and such other authorities as the judge never had heard of before. The outcome was that we were freed, and each of us received $2.00 for the work performed while prisoners. By that time, we had lost all interest in the beauties of Florida and returned to New Orleans with Colonel Holden.

Somehow, the Colonel got the life story of Three Finger Joe. The Colonel learned that Joe ran away from home at the age of seventeen because he was afraid of the consequences of a childish prank. The prank resulted in the death of a neighbor's calf. The Colonel had us work until we had $80.00. We then talked Joe into returning home just before Christmas.

I could tell endless other stories about our lives in the jungle. I could tell you how the Colonel did finance Spotty's return to his aged mother with enough capital to have him buy her a small parcel of land in Chihuahua County. Half of the stories of the Colonel's good deeds would fill this book.

Oh, yes, when we hoboes were on the road, we did go hungry at times. We were rained out occasionally, but what of it? We knew for sure that eventually we would eat, and we knew that rains would be followed by sunshine. And if it were very cold in the hobo jungles, there were always clean jails with friendly jailors and a handy judge willing to send us into one of them for a day or two for vagrancy. I myself was in the jail of Aholt, Missouri, for three days to escape the cold. I had a grand time playing with the two nice kids of the jailor, in return for which I was fed tasty pumpkin pies by the jailor's wife.

Hoboes worked as little as was necessary. But when the Colonel somehow heard about a widow near Hammond, Louisiana, having a very hard time of it, we did not hesitate to invade her farm and pick her cotton free of charge except for our food while working.

And Two Bits, Spotty, Three Finger, and I had a glorious time in New Orleans. The four of us were steady customers in an open-air dance hall, called Fern-Dance, and only about $.20 was needed to have a very pleasant evening of it. The dance cost $.05, and $.10 was needed to buy two ice cream sodas, one for the girl and one for myself. The bushes around the dance floor were of excellent service. Did I say that we shared everything in that strange hobo society? The four of us even shared occasional sweethearts.

But even if we were broke, I enjoyed watching the performance of the dancers immensely. The clientele of the place was composed of sailors of all races, including Negroes, Mexicans, Chinese Kanakas,[4] whites, and some local youngsters of doubtful occupations. It was a sight to see the manager at the end of each dance as he struggled to separate Kanaka sailors from their white girls who happened to be in the midst of—sexual enjoyment. Alas, nothing is permanent in this world, especially youth and pleasure.

When both Three Finger and Spotty left us, Two Bits Mike and I decided to look over New York. We left the old folks at New Orleans and took freights, at which we were experts, for a quick trip. After a few days during which we supported ourselves in style by Mike's dice and stayed at the Hotel de Gink, our wanderings took us to Battery Place. There, we saw a poster recruiting workers for the Panama Canal. The palm trees, the girls in sarongs,[5] and the picturesque tropical settings were very tempting. We applied for employment. Two Bits was accepted. I was turned down. It took me two hours to convince Two Bits that he had to accept the job, that he had to leave me. After spending the last evening together, we parted.

The Hotel de Gink was the most friendly place in New York. It was maintained on the Bowery for hoboes and lodgings were free. A plate was at the entrance for donations. No one watched the size of donations of the quests, and nickels were mixed with dollars and even $10.00 bills, for many an ex-hobo stepped in occasionally to drop a good-sized bill on the plate. In later years when I was in New York, I never failed to visit the de Gink and to contribute a few dollars. I was not alone by any means. The hotel was closed and dismantled about twenty years ago.

While the Hotel de Gink was virtually a home for me, it became a strange and cold place after the departure of Two Bits Mike. After leaving him at the recruiting office, I sat alone on the bench at the Battery Place. I felt that I never could return to either the de Gink or the hobo jungles.

CHAPTER 8

Romance

★

Noting the sign "Hungarian Home" on a building and feeling terribly alone, I felt that I had to turn once more to Hungarians for companionship and for moral support. The Hungarian Home was maintained jointly by the Hungarian Government and an American-Hungarian Society, and the Home offered shelter and food for newly-arrived immigrants at a modest price. It also furnished a free-of-charge home for three days for Hungarians who were broke if the Hungarian Consul's Office furnished a letter of recommendation.

I went to the Seventeenth Street Hungarian Consul's Office for such recommendation, hoping that during the three days in the Home I could look for a job. That was the purpose of the assistance. I was disappointed in this hope due to the stupid management of the house. One has to look for work early in the morning, but we were not permitted to leave the establishment before breakfast. Free guests had to clean up the place after breakfast. Therefore, it was about 10:00 A.M. when free guests were permitted to look for work. But, at any rate, I had lodging and food for the three days and, on the third day, I also obtained a job when another Hungarian Home's one-armed man came to the Hungarian Home's office and asked for five men. The man was the Hungarian foreman of a stone quarry at Tompkins Cove, New York, and when in need of men, he always turned to the Hungarian Home.

When the stone quarry foreman had made his needs known, the Hungarian Home manager called me to his office along with four other young Hungarians. But Mr. Wargay, the foreman, initially refused to accept me for the same reason that I was not taken to the Panama Canal. I was very skinny, just skin and bones. But the home manager saw the tears in my eyes and told Mr. Wargay that unless I were accepted, the Hungarian Home would never again furnish men for the stone quarry. I was accepted.

Mr. Wargay was sorry to have hurt me and to make amends, he talked to me very friendly on the way to the quarry. He told me that he would secure me the best Hungarian board with his own sister. I was touched by the unexpected kindness. When Mr. Wargay's brother-in-law, Mr. Szabo, took me to his home, I told Mr. Szabo frankly that I hesitated to go in because I would take lice into his home.

Mrs. Szabo was called out, and the three of us went into the wash shanty. I undressed completely. Mr. Szabo cut my hair with a number oo machine. I was given hot water with plenty of soap, new underwear and an overall, and my old clothes were burned. And no one knew anything about the reason of my change of clothes before I went into the house to join the other boarders. They were told by Mr. Szabo that I wanted to take a much needed bath. One has to be down and out to be able to measure the kindness of the majority of the people.

Mrs. Szabo was the prototype of the immigrant woman, and her life was very much like that of hundreds of thousands of others from Eastern and Southern Europe. When Mr. Szabo was released from the Austro-Hungarian Army at the age of twenty-three, he set sail for the United States. He promised his sweetheart that he would send her a steamship ticket as soon as possible and that upon her arrival in the United States, they would be married. Five months later, at the age of eighteen years old, Mrs. Szabo and her man were married in the Hungarian Protestant Church in New York. Upon Mrs. Szabo's arrival at Tompkins Cove, she took charge of the boarding house purchased previously by her new husband from a Hungarian family about to return to the Old Country. There were then seventeen boarders working in the quarry, all of them husky young men who were hard workers and hard eaters.

The young bride had to cook for nineteen people from the very first day, and it was not an easy task to familiarize herself with the American food, such as the salted ham and pork, cold-storage eggs, canned milk, etc. But she was pleased with the abundance of meat, white flour (used once or twice a year for cakes in Hungary), and canned milk. Her neighbors had to teach her to brew coffee, an unknown art to the Hungarian peasant women. Mrs. Szabo had to make six enormous breads twice a week in the outdoor oven built for that purpose, and she had to wash the clothes for nineteen people in the pre-washing machine age. Water had to be carried from the well which was about two hundred yards from her home. An average of thirty pails full of water a day were needed, except on wash days, when much more was needed. However, some water was stored before wash days by Mr. Szabo.

Mrs. Szabo had to get up at 5:00 A.M. except on days when she baked bread or washed. She was up at 3:30 A.M. on baking and washing days in order to provide breakfast to the boarders. Then came the children, the cleaning of the house, the buying of meat and groceries, the caring for the few pigs owned by the Szabos, more cooking, tucking up the children for bed, and by 8:00 P.M. or 9:00 P.M., Mrs. Szabo collapsed on her bed, dead tired. I left out small items such as sewing, taking care per chance of a sick child, etc., for these small items did not occur every day. But they occurred often enough to unbalance Mrs. Szabo's daily tasks.

About once a year, Mrs. Szabo was confined for two days following childbirth, except for once when she was blessed by twins and had to rest for three days. Mr. Szabo had to take over during confinements, and this left much to be done by Mrs. Szabo upon her third day. The only variation was one more child or two or three more or less boarders. Life and work for the Szabos went on like this for 365 days a year, except on Sundays, when the men did not work and Mrs. Szabo arose at 6:00 A.M. instead of 5:00 A.M.

During Mrs. Szabo's first six years in the United States, she never left Tompkins Cove. But on the sixth anniversary of her marriage, she entrusted her children and her boarders to her sister-in-law for a day while she went to the Hungarian church in New York with her husband. She also did some shopping for the children but returned the same evening. Mrs. Szabo stayed put in Tompkins Cove for another three years. But when the quarry was definitely closed, the Szabos moved to Red Jacket, Michigan, where Mr. Szabo's brother was working in the copper mine.

Years later, when I met the Szabos in their new Michigan home, Mrs. Szabo had eleven children and thirteen boarders. Her life was much the same as it had been in Tompkins Cove, except that the cruel winter cold of the Upper Peninsula of Michigan was especially hard on washdays. Although Mrs. Szabo had fewer boarders to care for, the routine of her house was much more complicated because men in the copper mine were working in three shifts and had to be fed at different times. Also, some of the children went to high school and commuted by street cars to Calumet, Michigan.[1] This meant additional tasks for their mother.

But Mrs. Szabo smilingly assured me that her life was somewhat easier because two of her growing daughters were helping her after school. And, about once a month, Mrs. Szabo went to the town. Assisted and interrupted as she was in the stores by the children, she was able to purchase necessities cheaper than in the company store.

As a result of both husband's and wife's hard work, the Szabos always saved money. The first saving of $1,700.00 was lost when the Rosett Bank in New York failed and, from then on, they sent their savings home. When they had $2,000.00 in the Bank of Hungary, they instructed relatives to buy a certain parcel of fourteen acres of land in their village and, within a few months, the Szabos paid the balance of $400.00 on the purchase price of $2,400.00. The Szabos had $900.00 in the Bank of Hungary at the outbreak of the First World War. They planned to return to Hungary to build a new home when the savings reached $2,000.00. However, when the war ended, their Transylvania village was included in Romania. Realizing then that their village couldn't be home for them anymore, the Szabos decided to remain in America for good.

The Szabos' decision to remain in America was arrived at by much heartbreak, mostly for the children's sake. The Szabos left Red Jacket and purchased a farm near Flint, Michigan. During World War i, they had saved over $3,000.00, enough to make a substantial down payment. The post-war inflation took care of their savings in the Old Country, and their farm at home was used by and later transferred to close relatives as a gift.

Upon returning from overseas in 1947, I visited on the Szabos' farm. They were prosperous, but not wealthy, for the education of some of the children was expensive. Four of Mrs. Szabo's sons, one of the daughters, and two grandsons were members of the United States military. One Szabo boy is buried at Guadalcanal, another, the attorney, returned a major from World War ii. Elizabeth, the nurse, returned a captain, and the other two boys, both farmers, became sergeants. None of them applied for deferments. They were raised to live in and appreciate America. In July, 1947, I found Mrs. Szabo with a hoe in her hands, still a typical immigrant woman taking care of her house, her flower garden, and her vegetables.

As I watched Mrs. Szabo working barefooted, I could not help but think of the 100 percent American ladies, some of whom often smiled with superiority and with ridicule and contempt if they happened to see the Mrs. Szabos and women like her with their ill-fitting clothes and work-thickened ankles. The 100 percent ladies were usually on their way to a bridge party or perhaps to a lecture about the necessity of shutting America's doors completely in the face of undesirable, non-Anglo-Saxon immigrants.

The morning after I first arrived at the Szabos' Tompkins Cove home, I went to work. While the work was quite hard at first in my weakened condition, it was nothing compared to work in the steel mills or mines. We were working in the open with the beautiful scenery in view — Tompkins Cove is right on the Hudson River. All that I had to do was crush stone with a hammer.

I was happy once more in the Szabo home. The room had its drawbacks, for I slept in the attic on a home-made bed that was originally manufactured for the fourteen- year-old boy of the house. It was a box, with straw mattress, and it was about six inches too short for me. I could not pull up my feet, for the box was too narrow for that. The only other alternative was to leave six inches of my feet hanging over with the board cutting into my ankle. But when one works in a quarry, little things like this do not disturb his sleep.

To return the kindness of the Szabos, I started to teach their three children to read and write Hungarian. The parents were happy, the rest of the boarders were friendly, and I soon also was teaching two illiterate boarders. Other boarders requested me to teach them to divide and multiply, and the Szabo home became a regular night school. Finally, I had reached a place where I was not considered an outsider, where I was not looked down on, and where I was the popular equal of my associates. Too bad it did not last long.

A few weeks later, the quarry was shut down. I refused the free board willingly offered by the Szabos for teaching their children, for I considered it charity, and my self-respect would be regained when I was once more a decent working man. I left the kind folks and did not meet them for over four years when, as a peddler, I found them at a copper mine in the Upper Peninsula of Michigan.

I returned to New York, but before looking up my old hangouts, I rented a dark hole that was called a furnished room for $1.00 per week. I started to look for work. Through my landlady, I became acquainted with a Swabian Hungarian[2] painter and wallpaper hanger and became an apprentice to him at $4.00 per week. The job looked good to me, and $4.00 was enough to carry me if I continued to eat free lunches, wash my own laundry, and pay $1.00 weekly for my room. Skilled men were earning as much as $16.00–$18.00 per week, and I was hopeful that within a few months I would learn the painting and wallpaper hanging trade. The work did not look hard, but it was not easy. Washing of the calcimine[3] or scraping off the paper of a high ceiling is hard on the wrist, especially when standing on a bending scaffold. Hanging on paper without wrinkles is an art hard to acquire. The worst part of the bargain was that the boss did not have steady work, and, of course, I was not paid on idle days. Nevertheless, I was happy, for I not only had work once more, but I also was learning a permanent trade. I was rather proud when, wearing my white overalls, I first visited Kossuth Hall and Rakoczi Hall and found the *SS Carpathia* bunch in the Kossuth Hall. My joy at the reunion was somewhat mitigated when I was touched for several nickels.

Within a few weeks, my pay was increased to $6.00. This meant that I really averaged about $4.00 to $4.50 a day, for we seldom worked over four days a week. But even this slender income enabled me to acquire some clothing. On Saturday nights, I went to Huston Street where innumerable pushcarts were offering all sorts of goods. I hardly ever paid more than $.15–$.18 for a shirt, or $.09–$.11 for an undershirt or a pair of drawers. A pair of socks did not cost me over $.03. I was very proud and very happy on the day when I purchased a fine-looking suit of clothes for $2.00 in a second-hand store on Seventh Avenue. The fact is that a good second-hand suit is far better than a cheap new suit. Workers and poor folks wear out their suits completely, and no cheap cloth lasts long enough to reach a second-hand store. Butlers and other servants were re-selling the clothes of their masters when the masters tired of wearing something. Knowledge of this fact was very useful, even when I was a clerk and later a "newspaperman."

Besides washing my own laundry, I shined my own shoes and wore collars which were sold two for $.05 and usable on both sides. Hence, two collars did for four weeks, and I looked clean. I took my meals mostly at free lunch bars, but twice a week I took regular meals at $.15 a throw.

Thursday was a red letter day in my life then for about six girls spent most of their time on Thursday in the furnished room house where I lived. They were servants. When they were between jobs, they roomed in our house, and they kept some of their stuff there permanently. They always came to our room house on their off days. I was twenty years old, healthy, comparatively well-off, and had lots of fun with the girls. The fun was had until one of them told me that we were to get married and with no nonsense about it.

The "shot-gun" marriage is an old and accepted American institution, but immigrants were better acquainted with "court" marriages. All an immigrant girl needed to do in those days was to go to a judge, tell the name of the unwilling man, and the unwilling man was given a choice of immediate marriage or jail for an indefinite period. Immigrant girls learned surprisingly quickly that ladies were protected in America. Since the girls served in American homes, they also were forced to learn English much faster than men. The girls knew their ways about quite well. And the usually penniless immigrant, unable to express himself, had little chance before the judge in such cases. I felt that the ground had fallen from under me.

Although I had not written to Icza, except the one letter from Portsmouth, I was still dreaming about my return to Hungary, and I was still in love with Icza. I tried to beg off from the servant girl by pointing out that I was making only about $4.00 weekly and that she was about ten years older than I. But I could not get to first base. The girl told me that she would leave her job, accept employment in a cigar factory where she could make $9.00–$10.00 weekly, and that we could live nicely on $12.00 a week. Further, she had $70.00 saved up, enough to furnish our home. I was panic stricken, and I did not know that I could escape her clutches by leaving New York. All I knew about the girl was that her name was Mary, she was about ten years older than I, and she had accepted my invitation to come into my room at our second meeting. She gave me one week to think over the marriage proposition. It was the unhappiest week of my life.

A week later, I told Mary that I was willing to marry her, but I requested another week's time to change my religion. Mary was Catholic, and I was still a Jew. Even when I was twelve years old and suffered all the consequences of the ghetto built around me by my Jewish father, I knew that I never would father Jewish children. Jewish readers may read this with contempt and call me a coward. I can return the compliment to many of them.

There are many Jews who have no more to do with the Jewish faith than I did and who would be happy to escape the discrimination practiced against them, even in America. But they suffer in silence, afraid of the prices we have to pay for complete assimilations. The unhappy Jews must suffer the alleged contempt of their relatives and friends and the reluctance of the Gentiles to accept even converted Jews. And so, the unhappy Jews remain Jews, leaving their children with the same handicaps that their parents inherited from their unhappy parents. Many of my Jewish friends admitted to me that were it not for these handicaps, they would be only too happy to part with their religion, if there were indeed any religion connected with being Jews.

Of course, assimilation is a sort of a privilege, and it has to be paid for. But all of us immigrants have to pay for the assimilation into the American communities. And some of us have to pay dearly. No so-called 100 percent American family greets with joy a prospective son-in-law with Hungarian, Greek, or Polish parents, even if the young man were born and raised in America. It does not matter how desirable the man may be otherwise. A good many native-born American girls leave their homes in tears because the unhappy parents are unwilling to bless their daughters' marriages to Dagoes, Hunkies, and Hunyaks. But the offspring of the mixed marriages are usually greeted with love even by the 100 percent American grandparents. When the offspring grow up, they know nothing about the price paid for the assimilation of their parents and are even proud to acknowledge their foreign forefathers.

Cruel religious and ethnic discrimination is felt in professions and, to a lesser extent, in business as well. And many hundreds of thousands of foreign-born men, or men born of foreign parents, suffer untold and undeserved miseries because of the discrimination. World War II was a blessing for many of America's enlisted foreign-born men because the distance and the unknown future led them nearer to their parents again and because, as ex-GI's, the men lost much of their inferiority complex caused by the discrimination that they were subjected to at every step before they entered service. And the multitude of the foreign-born ex-GI's felt that they were 100 percent Americans after World War II and thus were ready to kick in the immigrant gate that they were unable to crush before if, occasionally, they found the gate still closed in front of them. I know this because I am one of those foreign-born ex-GI's.

Reverend Kovach, the then Hungarian Catholic Priest of New York, was glad to receive me into the Catholic faith, but he told me that I needed a godfather and a godmother. When I suggested Reverend Kovach's janitor and his wife, I realized that Reverend Kovach wanted to make a show of the baptismal ceremony. He suggested the Hungarian Consul for godfather, and a Mrs. Varhelyi for godmother. Later, I discovered that he was on the outs with the consul and was looking for a chance to make up. Mrs. Varhelyi was the wife of a poor shoemaker who was the president of a church society supporting Reverend Kovach's church. I did not care about any of this for I felt that in my desperate situation, nothing mattered anymore.

I took Mary to Father Kovach to prove that I needed three weeks' time. Mary did not mind this for the delay meant a church wedding—something way beyond her own expectations. Doctor Arthur Kosa, a Hungarian friend of the consul, was requested to act as a go-between and to induce the Hungarian Consul to accept the doubtful honor of becoming my godfather. The consul accepted the honor. Baron Henry O'Conner agreed, and the wedding day was set.

Doctor Kosa was one of the finest men whom I had the privilege to know, and he assured me that when I was related to the consul, my struggles would be over. There was a large bank in New York owned by the Hungarian government, and the consul had his eye on a job there. But I knew that I would never accept anything from my prospective godfather because I did not wish to profit by my religious conversion.

About ten days before the date of my marriage, I had one of several talks with Doctor Kosa who had grown fond of me and whom I visited almost daily. I told him, frankly, that I was very unhappy about the marriage that was to be forced on me. Dr. Kosa started to laugh until his tears were flowing and assured me that he could end my misery. He said that he couldn't divulge the confidence of his client, but that I would be quite safe in telling Mary that the marriage was off because I had a talk with Doctor Kosa. And, if Mary did not like what I told her, she could lump it.

One can't imagine how happy that I was. I felt like a condemned man leaving the death chamber just before his execution. No half day was ever as long as that afternoon for Mary was meeting me nightly. When Mary arrived that evening, she went into a terrible rage. She did not succeed in scratching my eyes, but I did have a few marks on my face the next Sunday when I was baptized. But, I was a free man once more.

I met my godfather in church. I did not meet him before my baptism, and I've never met him since my baptism. Doctor Kosa urged me to call on my godfather because that would be the decent thing to do. But I was afraid that my godfather would insist on assisting me, and even though one may refuse the kindness of one's godfather, one can't very well refuse the wish of a consul.

Mrs. Varhelyi was a kind old lady. She kept boarders and invited me to move to her home when my room rent was up. This I did gladly. Following the incident with Mary, I wanted to move, anyhow. Three weeks later, my painting and paper hanging boss died, and I was without a job once more. The unhappy hunt for a job, any kind of a job, was on again.

The Varhelyis' board was excellent board with rich Hungarian food for $5.00 per week. While the Varhelyis were happy and somewhat proud to keep me, and while they assured me that they considered me their son and that I need not worry about payments, I marked up $5.00 to the Varhelyis each Saturday in my little notebook. The Varhelyis had no children of their own.

Not having much else to do, I started to organize Mr. Varhelyi's business. He had a small shoe repair shop on East Sixth Street, and I induced him to buy new shoes with his $300.00 savings. Mr. Varhelyi had been President of the "First Hungarian King St. Stephen Roman and Greek Catholic Hungarian Sick and Church Society for Men and Women" (believe it or not, that was the name of the society, and there were some with even fancier names), and the society members visited his shop often. Some started to buy shoes from him. Even when Mr. Varhelyi realized that he made $.50 or $.75 on a pair of new shoes, without working, it took me days to induce him to expand and to request more goods on credit. I had to go with him to the wholesale house of Powell & Campbell, I had to be the spokesman, and Mr. Varhelyi was surprised when a credit of $300.00 was granted to him. We built shelves at his shop, and I forced him into buying a small showcase for the street. It was shined religiously every morning by Mrs. Varhelyi, and I daily changed the shoes in it.

The Varhelyis were very proud and reassured me almost daily that I did not need to worry about the board and that I was a welcome member of the family. But I did have a family in the Old Country, and I knew that they would be very unhappy about my changing my religion. I had advised my oldest brother of the change immediately, but that very unhappy fact was never mentioned in letters that I received from my folks.

I kept on looking for work and months later, when I owed over $100.00 to the Varhelyis, I did find work in the McAdoo Tunnel.[4] The work at the tunnel was a terrible job. I was assigned to the night shift. My job was to carry sacks of cement from the elevator, as the sacks were lowered into the mouth of the tunnel to a small train that took the sacks to the working tunnel face. All of us were working in compressed air that was to partially support the roof. I had to rush, even run, with sacks of cement on my shoulder. I sweated excessively and, within a few days, both my shoulders were full of sores.

The Varhelyis begged me to quit the job and finally, Mrs. Varhelyi once refused to wake me up at 11:00 P.M. to start for Courtland Street. Because I was two hours late for my job, I was fired. Ever since then I feel some pride whenever I travel through the tunnel.

Mrs. Varhelyi tried to refuse to take my money. I had to threaten the Varhelyis that I would take same and pay their account at Powell & Campbell in order to convince them to accept my payment.

I was again out of a job, and I was very unhappy. The amount owed in my book grew by leaps and bounds, and I spent many sleepless nights trying to figure out the number of months or years it would take me to pay my debt. I would have liked to return to coal mining, to scrubbing saloons, to sleeping in the park, anything to put a stop to the growth of the account. But, I was a coward.

After the Ulmer experience, I was afraid that until I could pay something on account, the Varhelyis also would believe that I left them in order to escape my debt. This fear alone kept me with the Varhelyis. I always was looking for but never finding a job, and my indebtedness ran up to the hundreds of dollars.

Of course, I entered the same social circles the Varhelyis lived in by joining the church, the church society, and other organizations in which the Varhelyis were involved. As President of the St. Stephen Society, it was up to Mr. Varhelyi to see that his society prospered and that its membership of about four hundred and its funds grew. Picnics and amateur shows were held often, and I was a great help in organizing them. Partly to re-pay the Varhelyis' kindness and partly to pass time which was heavy on my hands, I took charge of the show activities and put on an amateur show almost every week at the St. Stephen's Hall. At other times, I was only the director. At times, I took small parts in the plays, parts too small to be accepted by more ambitious actors. I also entered church politics in the Catholic congregation. About a year after Reverend Kovach baptized me, his flock tried to get rid of Reverend Kovach. This was not an unusual matter in foreign circles in those days. Finally, Reverend Kovach got tired of the bickering of the people and arranged with the respective bishops to change places with the Hungarian Priest of South Bend, Indiana.

We did not care for the South Bend Priest who quarreled with his own people, and we decided to close in the Fourteenth Street Church from him. I also was requested by Mr. Varhelyi to write a pamphlet enumerating Reverend Kovach's faults, and one thousand copies of the pamphlet were sent to South Bend to be distributed there. The South Bend people returned the favor in kind.

When the South Bend Priest arrived to us, he found the church closed, and the people in front of the church received him with jeers, water bombs, and rotten eggs. He left by the same taxi with which he came and returned shortly with four policemen. The police dispersed the mob, opened the church, and the priest took possession.

A committee, including me, went to the police station for an explanation. We were told by the police captain that the church was recorded as the property of the diocese and, for all practical purposes, the bishop was its owner. Until then, we were not aware of this and even then we considered it very funny that a church built at the expenses of the people should be the property of the diocese, without any rights left for the people.

Such religious scandals and riots were every day occurrences in the Italian, Polish, Rumanian, and Hungarian congregations, and it took decades to have the immigrants understand the American system of religious life. And let it be said, no matter how unpleasant is the fact, that in most cases, priests and ministers—many of whom were dear friends of mine—were at fault. Especially in the early days of the immigration of particular nationality groups, few priests and ministers were induced to come to America by their religious zeal. More came because of some unpleasantness with higher church authorities or other personal reasons.

Churches had to be built, and since both priests and ministers were poorly paid, it was a great temptation for them to accept a little money from the contractor. This money acceptance was then discovered later by some member of the church. And if one such case were discovered, every priest or minister building a church was subjected to the same suspicion. Most of the priests and ministers were clean in every respect, but some of them were really bad. So bad that bishops were unable to put up with them, and some thus were kicked out of the Catholic Church. About 1900, an Independent Catholic Church was organized by such priests amongst the Poles, and priests were able in some cases to secure parishes of respectable sizes, mostly because of the rigid rule concerning ownership of churches.

Hungarian priests twice tried the same process as the Poles, but both efforts were unsuccessful. The Hungarian Protestant ministers received some financial assistance from Hungary until the First World War but, with the outbreak of the war, the ministers were left orphans, and the financial assistance was gone.

Most of these immigrant ministers took their organizations into the folds of the Reformed Church of America, but some joined the Presbyterians and one the Episcopalians. This divided the Protestant Hungarians and caused a religious warfare that lasted close to twenty years. Following the First World War, a few zealous and especially able young ministers came from Hungary and organized self-supporting Protestant churches in colonies already having such a church that belonged to some American organization. This intensified the bitter war amongst the Protestants.

As I worked with the St. Stephen Society and in later years with much larger organizations, I could not help but see the weakness of the Hungarians who were applying democracy to their social life with a vengeance. As a result of not having any say-so in their public affairs in Hungary, they were not trained in the first requirement of a democratic form of life: self-discipline. Their newly-won freedom, their lack of training, and their just suspicion of their educated fellow Hungarians handicapped them fearfully in establishing and maintaining the institutions around which they lived their social lives. And I was fearful that this utter lack of self-discipline would be a tremendous handicap, even for the Hungarians of Hungary, in establishing a democracy now that they were rid of the old ruling class.

In 1910, there were about four hundred Hungarian societies scattered all over the country east of the Mississippi. Some were organized to assist the sick, some to bury the dead, some to assist churches. Some organized along religious faith, others to embrace Hungarians from a particular county of Hungary. In days of depression, many societies had to fold up because many of the members returned to Hungary or sought jobs elsewhere.

Many of the Hungarian societies tried to imitate American fraternal organizations by dressing their members in uniforms to march for holidays or funerals, and the gaudy uniforms always resembled that of the Austro-Hungarian army officers (a thing beyond the dreams of a Hungarian peasant in Hungary) with a few decorations added for good measure. The Hungarian press, in trying to encourage organized Hungarian life, encouraged the unnecessary extravagance, and the poor immigrants who were denied such pleasures in the Old Country were spending a month's earnings to be able to send home a picture of themselves in a colonel's uniform.

Flags were purchased by every Hungarian Society, and the flags were expensive, often costing $500.00–$600.00. Of course, each society's flag always had to be more expensive than the one owned by the rival society. Cabinets were designed and built for both the Hungarian and American flags. Hard-working Hungarian women who kept perhaps twenty boarders were eager to spend $100.00 for the honor of acting as sponsors of those flags. A flag sponsorship meant that a photograph would be sent to the sponsors' villages in the Old Country. The photograph showed a wide ribbon encircling the sponsors' stomachs and all the glories of the sponsors' day of fame. A clipping from the Hungarian newspapers was included with the photograph.

Many of the small Hungarian societies collapsed, and one society finally emerged to join the remaining societies into a national fraternal organization. But no sooner was the national society organized than lack of discipline led to its disruption. The minority of the society was still unable and unwilling to abide by the decisions of the majority, so the minority organized another society and the rivalry was on. A second, third, and a tenth society were organized, some by well-meaning groups, some on religious grounds, and a few by crooks. Many societies went broke because their leaders and members were equally ignorant of finances, and some went broke when their leaders skipped with the treasury. And even the societies which had weathered the storms of the early days—and the Hungarians now had four strong and healthy national fraternal organizations—had a hard sled for lack of financial knowledge and ignorance of mathematics.

I was present at one bi-annual convention of the strong and healthy American Sick Benefit and Life Insurance Company of Bridgepost, Connecticut. The convention was held in Woodbridge, New Jersey. The approximately two hundred delegates sent by their branches were very reluctant to increase membership fees by a few cents although they were shown clearly that the organization was bankrupt at the time. Finally, the delegates agreed to increase membership fees. The week-long convention was about to end when one of the delegates reminded the meeting that a rival organization was offering a $100.00 larger policy, and another delegate made a motion to increase the Bridgepost Company's policy by $200.00. Motion was adopted enthusiastically, and neither the newly-elected officers, nor the delegates, seemed to worry that the larger policy would need still higher membership fees.

It took many years and Hungarians had to lose many hundreds of thousands of dollars before the various American insurance departments were authorized to exercise some control over the Hungarian insurance companies and force the companies to reform their organizations on a healthy basis. And as far as I know, this also was the history in every other nationality group.

Educated men were not welcome in the insurance organizations for many of them were of doubtful character, and the rest of the organization members had to suffer the consequences of few educated members. The above-mentioned American Sick Benefit and Life Insurance Company even stipulated in its by-laws, until about twenty years ago, that no member could be an officer of that organization unless he previously engaged in physical work in the Old Country. Decent, educated men resented the discriminations and suspicions of the organizations. Even now, the weakness of the Hungarian-American institutions is due to the fact that educated Hungarians who are nationally respected for their achievements in the fields of arts and business are unwilling to acknowledge any relationship with their fellow countrymen and thus stay away from the Hungarian colonies.

CHAPTER 9

In the Show Business

★

WHILE STAGING SHOWS for the benefit of the St. Stephen Church and the St. Stephen Society, John Kocsel, one of my co-workers, and I had a bright idea. We decided to become producers for our own profits. In those days, there were some professional Hungarian actors who occasionally put on shows in the Webster Hall, the Progress Hall, or the Manhattan Casino, and there were also amateurs who invaded the show field.

John's and my decision was followed by immediate action. We organized an Amateur Club, secured actors, selected a play, and rented the Manhattan Lyceum for the performance which — as usual in the time — was to be connected with a dance.

A newspaper was published to properly advertise our show with hopes that if we made money, we would publish the newspaper permanently. Newspaper advertisements were secured from Hungarian business people by Mrs. Varhelyi and her friends. These Hungarian business people and professionals often were solicited by charitable organizations and church publications and, as a result of their contributions, the first and only issue of the *Trail-Blazer* was a huge financial success. John Kocsel and I made a profit of over $60.00. Excluding this profit, money was advanced for necessary show expenses by Varhelyi and the widowed mother of Kocsel. John and I then decided to stage a show that would be remembered by the Hungarian colony.

Appropriate show costumes were rented and, since the first act of the play was staged in a garden, live rose bushes were secured in the middle of winter at high prices. An active fountain was built in the middle of the stage with very unpleasant consequences. The act was long, the wash tub under the stage failed to hold the water from the fountain, and several pieces of props stored under the stage were damaged. We had quite a fuss with the owner of the hall about it. The second and third acts were staged in different rooms, and I—experienced paper hanger—re-papered the room between acts to the amazement of the audience.

But while the show was good and far better than even those produced by professionals, John and I knew during the performance that we were bankrupt and that it would be our last production. An attendance of four hundred people was needed to have us break even, and an hour before the show, New York was visited by an unprecedented downpour. About two hundred tickets had been sold and their owners came, but no one else risked the trip in the rain in those pre-automobile days. The producers as well as the backers were downhearted during the show, and we were a gloomy bunch when the after-show dance started with our facing a deficit of about $100.00.

But a slightly drunk blacksmith and a sharp trick on my part saved the day. The dance was connected with mock marriages which meant that any young man could marry a girl for $.10, with the man receiving a kiss and the girl an engraved certificate of the event. Two prizes were offered for the girl with the most marriage certificates. The first prize was a sugar bowl with twelve spoons hanging around the bowl. I paid $1.25 for the sugar bowl. The second prize was a far more valuable basket of flowers which was presented to us by the florist who sold us the rose bushes.

As soon as I noticed that the blacksmith was trying to win the first prize for his girl, I gave $50.00 to one of my friends and instructed him to marry our leading lady as often as the blacksmith married his sweetheart. At the end of the marriage race, the blacksmith was permitted to win the first prize. But, winning cost him close to $70.00. Including the income from other weddings, John and I managed to wipe out the threatening deficit. But my days as a producer were definitely over.

During my stay of over forty years in the United States, I must have committed other sins and crimes, but I always will remember and never will forgive myself for cheating the blacksmith. My only defense is that both the Varhelyis and Mrs. Kocsel were very poor and could ill afford the deficit.

In those days of Depression, there was one firm, the United States Coal and Coke Company of Gary, West Virginia, that constantly advertised for men. Following our inglorious finish to our careers as producers and publishers, I induced John to try coal mining.

My plan was opposed both by the Varhelyis and Mrs. Kocsel, and my suggestion of riding the freight trains was even more opposed. John and I were given $10.00 each, and we took the boat of the Old Dominion Line to Norfolk. We told the old folks that the rest of the money would be enough to pay for train fare from Norfolk to Gary, which was not the case.

John and I spent the night on the boat deck for cabins would have cost extra. I then got acquainted with Mr. Holezer, another Hungarian passenger who was trying to establish a Hungarian farming colony around Norfolk. The colony was to be called Kossuthville. Mr. Holezer told me that his efforts were about to fail because the Hungarian newspapers were opposed to the farming scheme. The newspapers were opposed mostly because Holezer did not have enough money to advertise, or, as he bluntly stated, to bribe the newspapers. Settling Hungarians on farms was so logical that I could not believe him, but later I knew that he had told the truth.

Hungarian colony publishers were then struggling. They were not bound by the ethics of the journalism profession and, at times, they failed to draw the line sharply between legitimate advertising and blackmail. Newspapers were started without capital, publishers and other leaders were jealous of each other, constructive enterprises were attacked mercilessly if they failed to come across with what publishers considered proper amounts for advertisements, and never-do-wells resented any seeming success of the legitimate successful publishers.

Representatives of the Hungarian government were the highest authorities in the Hungarian colonies. These representatives had outstanding prestige, their hints were the laws of the colonies, and they were opposed to Hungarians' settling permanently in America. The approximately $100,000.00 sent home by Hungarian immigrants was needed by the Hungarian government to balance their budget.

And, while German and Dutch bankers invaded America and purchased large tracts of land to offer to their immigrants—tracts such as the still-flourishing Hungarian colonies of Michigan, Washington, and Texas—the Hungarian government and its representatives steered their own immigrants to steel mills and coal mines, thus sentencing them to hard labor and to the permanent fate of industrial proletarianism. Many Hungarians returned to the land and farming individually when, after the First World

War, they had to reconcile themselves to remain in America for good rather than return to Czechoslovakia or Romania. They became excellent farmers.

Some Hungarians returned to the Old Country to enjoy the piece of land they had purchased with their first earnings or the new home they had their relatives build for them. They invariably returned to the United States—broke. In spite of the nostalgia after their old village, they could not feel themselves at home anymore in the primitive settings because, in spite of themselves, they had become Americans. The atmosphere of a European police state choked them. Many of the immigrants went back to their native lands several times, but to no avail. Millions of immigrants of all nationalities and of every standard of education or wealth tried going back, but all returned to America. Even if they never learned English, the melting pot was effective.

When John and I arrived at Norfolk, we took a freight train and within three days, we arrived at Welch, West Virginia, the junction for the Gary branch of the railroad, about four miles from Gary. It was Sunday. We washed up in a hotel, walked the four miles to Gary, and arrived there about 2:00 P.M.

We were turned down in the first two Hungarian boarding houses where we sought board, but we were promised board in the third house by a young Romanian-Hungarian couple who had only two boarders besides us. The couple were newcomers and new in the boarding business. John and I were afraid to ask for food and were very willing when the young couple invited us to go for an ice cream, for ice cream also was food. It was a very warm day, half of the ice cream melted, and the young woman said the "sauce" tasted especially good. The sauce was the liquefied ice cream. Doubtless, the lady later became familiar with the various syrups, but I had to smile at her naiveté then. Her remarks reminded me of my first days in Portsmouth.

The boarding house beds had not been slept in for some time, the "linen" was covered by dust, and the supper was very bad. I had quite a time with John Kocsel about both.

Next day, we applied for work, but the surroundings and the whole atmosphere of the camp were not conductive to very bright hopes. The tremendous labor force was recruited from every corner of the country. There were men of every nationality with a good sprinkling of colored men. Dozens arrived and left every day. Drifters and floaters went through Gary by the thousands, and it took years to organize the excellent labor force of several thousand coal miners.

Gary was full of armed guards furnished by the infamous Baldwin-Felts Agency[1] of Bluefield, West Virginia, and the guards saw to it that the word of Mr. O'Toole, the camp superintendent, was law on the camp. If any man displeased Mr. O'Toole, the man had to move within twenty-four hours whether he had a place to move to or not and whether he had twelve children or not. It was a paternalistic, strong-arm rule, often ruthless, but I have to say that O'Toole — who much later became my friend — was always fair according to his own life. Kocsel and I were led to the mine by a foreman just at the time when most of the miners were blasting.

John was scared of the blasts and of the thick smoke coming out to the main entry. He turned around and ran as fast as anyone can run in a five foot high mine. I had to promise him in New York that if he couldn't stand the mine, I also would return to New York with him. I had no choice but to leave Gary, and I have to confess that the choice was not hard. I was spoiled by the Varhelyis' good home and food and by the idle days, and I left the Gary mine quite willingly.

Prior to returning to New York, John and I tried Pittsburgh where Varhelyi's brother was foreman in a wire mill. We had a very unpleasant incident on a freight train, an incident almost unheard of in hobo circles. While we were standing between two gondolas loaded with coal, the brakeman came around and hollered down to us. He requested a quarter from each of us. We refused indignantly, and the brakeman started to bombard us with coal. And as the lumps grew larger and larger, and as he made several hits, we realized that he meant business and forked over two quarters. But I was sorely tempted to kick him off the train when he stepped down from the top of the coal to collect. Only the presence of Kocsel saved the brakeman.

When John and I reached Pittsburgh, Varhelyi's brother was out of a job, so we proceeded to New York. After days of alternately riding freight trains and walking, John and I came to a then very strange sight. Hundreds of people were working on construction — building streets and erecting buildings. John was anxious to reach New York, but I induced him to let us look for work on some of the construction. We were promised employment digging ditches. I wanted to earn at least enough money to return the $10.00 to Varhelyi.

John and I also obtained board in one of the Hungarian boarding houses after having been turned down by several others, and I was amazed at the modern home. The houses were built of brick, each had a bathroom, and the yards around the houses were fenced. There were chicken coops and other out houses in every yard. This standard of living was something I never had seen at the disposal of immigrants. The New York tenant house I had lived in, as most of the other tenant houses on the lower East Side, furnished water in each hall for all the tenants on the respective floors and a toilet in the yard to every two tenants.

Since there were lots of Hungarians amongst the construction workers, I soon learned the story of this strange place, this ready-made plumb new town with a large store, post office, fire department, sidewalks, and landscaping. There was no visible source of income for the population except a large kiln. The strange place was Roebling, New Jersey, and I was told that there would be an immense wire mill in the town but the mill would be built only when the prospective labor force was taken care of by houses and other necessities. And according to plans, the kiln was the first building erected to furnish bricks for the houses. This Roebling community was one for the books, indeed. Whoever heard of such things in Europe as building a whole new town to accommodate a factory and its workers! And more, whoever heard of bathrooms for workers, even in America?

When John and I returned to the boarding house for supper, we were told by its owner that we were welcome to spend the night there but he had changed his mind about giving us permanent board. I was disappointed, but not surprised. The rest of the boarders coming home at the close of the shift must have heard about the two new men who arrived without suitcases or trunks and who allegedly came from New York but didn't look like real workers. The boarders must have been scared and likely they objected to John and me. They had no reason to believe that my and John's things would be sent after us. And in those lean days, there were many crooks going around who robbed not only boarders but who also robbed the boarding house Mrs. of the savings of the boarders.

Next morning, John and I started toward New York. I was tired of walking, I was penniless, and I was facing the unpromising immediate future. But I dreamt dreams. And in my dreams, I was building a town just like Roebling. Perhaps not quite as ambitious, not quite as large, but with workers' homes containing bathrooms and populated only by Hungarians. My hopeless status did not disturb me much in my dreams. I KNEW that sometime, somehow, the dream town would be built.

Madness?

Perhaps. But dreams are duty free, and youth's right is to dream. (It took me twelve years of dreaming but finally, to my joy and also to my deep and everlasting regret, the town of Himlerville, Kentucky, was built. And built it was with a bathroom for each coal miner.)

Both Mrs. Kocsel and the Varhelyis were happy to see John and me return. They hoped that the two of us were cured of the nonsense of coal mining for good.

CHAPTER 10

I Become a Newspaperman

★

SHORTLY AFTER MY RETURN to the Varhelyis, Dr. Kosa came to my rescue again. He secured me a job in the New York office of the Cleveland daily *Szabadság*, at a salary of $7.00 per week. My days as a laborer were over.

It was nip and tuck with getting the job, for one of my duties was to read the *Staats-Zeitung* every morning and wire the news items from or about Hungary to Cleveland in time for our evening newspaper. I did not know German. Dr. Kosa agreed to have me go to him every morning with the German newspaper, and he made up the wires. And for over a year, this man did that for me, even on the many mornings when I had to wake him from his sleep after a long night of house calls.

Dr. Kosa never failed to smile and to assure me that all of this was of no bother though Mrs. Kosa frowned on the arrangement from the beginning. To reciprocate Dr. Kosa's kindness, I placed small items about the wonderful recovery of some of his patients in the Cleveland daily *Szabadság*. The items also won over and pacified Mrs. Kosa.

The Varhelyis were very proud of my new career as a newspaperman, and, at first, so was I. And I found much pleasure in attending meetings, picnics, and banquets where I received the respect due a newspaperman. There were many such activities in the colony. I learned to enter the activities without handing over my press ticket which was always mailed to the newspaper but which often was sold by me to one of my friends at a reduced price. I had little work during the day except to write up the meetings of the previous evenings, accept a very few subscriptions and still fewer classified ads, and count the approximately five hundred copies that we received daily for the distributor.

National advertisers were handled by Mr. Hornyak, the manager of our branch. He was a very quiet and sour old man. There were only the two of us in the office, but there were days when we had nothing to say to each other.

My $7.00 per week salary provided me with $5.00 for board, $1.00 paid on my account with the Varhelyis, and $1.00 for clothing and incidental expenses. The knowledge that it would take years to pay my full indebtedness to the Varhelyis drove me crazy. I looked over my book almost daily, trying to figure out when and how fast my pay would advance and when and how I could liquidate my debt. I forgot to smile during what they call a man's best years!

I was a little over twenty years old. I wanted to have some fun, to mix with young men of my age, and, of course, to mix with girls. But when attending the numberless balls and picnics as a newspaperman, I had to stand at the wall lest my shinning pants invite attention. I was unable to return favors when one or the other guests honored the newspaperman by treating me to a glass of wine.

Once, I had an affair with a middle-aged woman, one of the fortune teller advertisers of our newspaper. Despite our relationship, she wanted to secure me for one of her clients, a beautiful Hungarian servant girl who had promised to pay the fortune teller a substantial sum for a husband. When the fortune teller left the servant girl and me alone in the fortune teller's home, "the thing" happened, no doubt because the girl took it for granted that the event would end in marriage. The girl invited me to call next Sunday afternoon, and I was happy with the anticipation of repeated performances.

I showed up early at the girl's home on the following Sunday before the family had dinner. The girl told me to wait outside for a half hour, at the end of which we would be free to go to a picnic. Realizing that one has to pay for everything in this life, and that I was not financially able to take her to a picnic, I skipped. Were you ever twenty years old, possessed a girl who was willing, and was then compelled to leave the girl for the lack of about $1.00? If not, it would be useless for me to describe my bitterness.

To boot my misfortune, my fortune teller friend requested me to give her an English-Hungarian Dictionary which was published and sold by our newspaper for $3.00. Not having the $3.00 to pay for the dictionary and not willing to steal one, I had to put an end to my relationship with the fortune teller.

Such was my life as a newspaperman. Meanwhile, to pass the time and to help Varhelyi, I induced him to open a regular shoe store on Avenue A. A good-sized store was rented and partitioned off into three rooms. The front was the store, the dark room next to the store served as the Varhelyis' bedroom, and the third room was the kitchen, dining room, repair shop, and my bedroom. One can easily imagine the various smells when I was about to retire in the evenings, but I did not mind.

I ordered two hundred empty shoe boxes that soon filled the upper parts of the store's home-made shelves, and I obtained larger credit in order to have some stock in the store. I soon realized that I would have to be the principal salesman and, luckily, most of the customers came in the evening when I was home. Varhelyi was unwilling to sell and even very reluctant to buy a pair of shoes if the pair were not serviceable. When a servant girl selected a pair of pumps currently in style, the shoemaker in Varhelyi revolted, and he tried to force a pair of good and strong shoes on the servant girl. Or, if a girl with size six feet happened to ask for size five and three-fourths shoes, Varhelyi was quick to enlighten her that she needed size six and double E at that. My policy was to hand over the size six double E shoes and assure the girl that she was misinformed about her size, for lo!, the size five shoes fit her eminently.

I spent my days between my newspaper job, the shoe store, and constant agony. When my pay was raised to $8.00 per week, I paid $2.00 on my account and occasionally saved enough to buy a second-hand suit of clothes out of the weekly $1.00 that I kept for myself. As a consequence of my profession, my circle of friends and acquaintances grew by leaps and bounds. Since people were unaware of my financial enslavement, I received hints and offers to marry. Varhelyi urged me to select a girl with some money and go into partnership with him, but the offer did not appeal to me. For one thing, I was still in love with Icza. For another thing, I did not wish to be a storekeeper.

I often met one of my friends, a Hungarian importer by the name of Szabolcsi. He and his brother were the owners of the oldest and most respectable Hungarian importing house and sold wholesale prayer books, novels, smoker's articles, and innumerable other small items. Once, I told

Szabolsci that he ought to go into the retail mail order business. A few weeks later, after talking this matter over with his brother, Szabolcsi offered me the job of organizing such a mail order branch for Szabolcsi.

I took up Szabolcsi on his offer and requested a week to serve notice on my newspaper boss. But by that time, I was so deeply attached to my unhappy boss that I was afraid to advise him of my contemplated leaving until the day before I left. I was pleasantly surprised when my boss told me that he knew about my leaving and that he was asked by Szabolcsi whether he had any objection to my leaving. Since my boss knew that he never could pay me more than $8.00 per week, he did not object. As I learned later from Szabolcsi, my boss had told Szabolcsi that I was honest as the light of day but very stupid and sometimes had nothing to say for days on days.

My new job began with a salary of $9.00 per week, and I duly paid $3.00 on my account each week. Shortly after, my pay was raised to $12.00 weekly. Out of this $12.00, I paid $5.00 for board, $5.00 on account, and finally, I had $2.00 each week for myself.

I started my new job by making a catalogue for the retail trade. I used every superlative expression known to the Hungarian language and wrote equally tempting text for the advertisements in the Hungarian newspapers. I objected to the steep prices in the catalogue, but Szabolcsi explained to me that he had to protect his wholesale trade, the large number of peddlers, and the Hungarian bankers and merchants who were his steady customers. The mark-up on imported articles for the wholesale trade was 100 percent, and another 100 percent was added for the retail trade.

The mail order department was a success from the first day. Hungarians in far-away places, such as mining towns, were not visited by peddlers, and they were glad to avail themselves of the mail order service. I mostly was left alone by my bosses, I managed the department myself, and I learned not only a lot about business amongst the immigrants, but also something about American business. Szabolcsi was a reliable businessman of the highest standing—by immigrants' standards.

The meerschaum[1] pipes sold by us were really meerschaum, but the "Adler" name of the best known and most reliable European meerschaum firm, known to every immigrant, was stamped into our third-grade pipes by ourselves. The silver watch chains worn by immigrants in those days were really silver, but instead of 950 fine, they were 800 fine. The Hungarian government's official silver stamp was stamped onto the chains in

Szabolcsi's store. The parts of earrings going into the holes on the ears were really imported 14 karat gold stamped officially by the Hungarian government's bureau. But the major parts of the earrings, the parts holding the stones, were of 10 karat gold and were attached to the earrings by a local goldsmith on Eldridge Street.

We sold hundreds of Columbia gramophones as were shown in our catalogue, but at least twice as many cheap, German-made machines resembling outward the far better and far more expensive American goods. I noticed that the Hungarian peddlers were buying gold jewelry from us, but were selling much gold-filled and even gold-rolled jewelry that they misrepresented as gold.

The prayer books in bindings decorated by bones were sold and represented as ivory-decorated books, even in Szabolcsi's catalogue. Cheap, third-grade American-made harmonicas and zithers were represented as imported goods and sold at terrific prices.

Immigrants were cheated and taken advantage of from the minute they decided to immigrate until their return—if they did return—to their Old Country. Hungarians were no exception. When applying for a passport, Hungarians usually were fleeced by the village notary who charged 20 Kronen while the official fee was 1 Kronen. Hungarians were disappointed by the accommodations on the steamers for instead of having the cabin shown by the steamship company leaflets, they had to put up with huge dormitories. Hungarians were fleeced by the Hungarian bankers when they sent their money home, and they very often were fleeced by entrusting their savings to private American bankers who failed. Hungarians were cheated by Hungarian Notaries Public when making an affidavit and were forced to pay $5.00 or $10.00 for a notary signature, instead of the official price of $0.25 per signature. Quite a number of the Mesdames of the boarding houses decided to change mates. When one of them left her husband, she usually took the savings of the boarders.

It is a wonder that in spite of innumerable disappointments and immeasurable losses, Hungarian immigrants were able to find their places in the American community. Class by class, the working immigrants were financially better off than native Americans.

At Szabolcsi's, I did my own buying of novelties, musical instruments, watches, and such other goods that we sold, aside from the imported articles, and I got acquainted with the wholesale houses of Canal Street and lower Broadway. I was amazed to discover that we could order cheap American-made watches from Europe for less than they were sold on the domestic market by the manufacturers. I also learned something about the high tariff and the business called "dumping." I was shown that a certain sewing machine, which was unloaded on the American public at the price of $40.00, was sold in Europe for the equivalent of $16.00 after paying duty there and transportation cost. I purchased from Paris a pair of well-known American shoes for two-thirds of the price at home.

My department of the business was so prosperous that my pay was increased within a few months to $15.00 weekly, and I was able to pay $7.00 on account each week. At the same time, I started to make money by serving summonses for one of my attorney friends, and I received $5.00 for each case. To my indescribable joy, my twenty-fifth birthday found me free of debt. I purchased a fine and nice blue suit for $15.00, the best obtainable in a large clothing store, and I felt like a million dollars.

Once, one of our customers in Houston Street phoned that he needed one hundred Hungarian Almanacs in a hurry. I didn't have a delivery boy handy, so I took over the large bundle myself. The Houston Street customer was a personal friend of mine, and we were located only seven blocks away. On the way to Houston Street, I met my Weiss cousin for the second time (the last), and he asked what I was doing. Since there was little love lost between us, I answered him shortly that I was taking this bundle to Deutsch Brothers and did not think anymore of the incident for some time. Within four weeks, I received a letter from my brother, the like of which I would not wish to my worst enemy. My brother's letter told me that my family members were aware of my style of life. They had learned that I became a drunkard and a bum and eked out a living by carrying packages. My family asked why, for God's sake, why did I not write home for passage money if I were down and out? They said that as hard as it likely would be to make a new man out of me, they were willing to try for the sake of my mother.

To understand this letter's message, one has to know that in Hungary it is a disgrace to carry a bundle. There are messengers for hire in Budapest, and bums are hired for such tasks in Hungarian small towns. This letter came when I was free of debt, and when I was a well-paid and respected member of the community where I lived. I was raving mad at my cousin, especially for stating that I was drunk for, as I have said before, to this day

I do not drink alcohol unless I am forced to. Even my brother remarked in the letter that my family was especially surprised by my being a drunkard.

Of course, I wrote a return letter to my brother and tried to explain my situation and my cousin's misunderstanding of my situation. To prove to my family that I was all right, I enclosed a 20 Kronen bill in the letter, the equivalent of $4.00. From then on, I mailed home such bills weekly in every one of my letters. I had no way of knowing, until much later, how much my family did believe me.

Aside from the letter incident, I was contented, interested in my work, and better off financially than ever before. But the desperate years of worry had left their mark on me for life. I became an introvert and a shy man, so much so that I often crossed the continent without speaking to anyone or crossed the ocean speaking only to the waiter, unless others made the first move to get acquainted.

Once, in the middle of November, 1913, the largest Hungarian private banker offered me a job as manager of his mail order department at a salary of $25.00 per week, the top salary obtainable in those days in office work. I was grateful for the compliment but turned down the banker's offer for our own mail order department did so well that I was sure Szabolcsi would pay me as much in the next year. I not only did well in Szabolcsi's business, I had become one of Szabolcsi's close friends. I helped him translate Hungarian plays, one of which, *The Tajfun (Typhoon)*, was produced by him on Broadway with measurable success. I did not have the slightest doubt that I would receive a raise.

When Szabolcsi failed to mention a raise for me at Christmas and the New Year and handed me the usual $15.00, I told him that I wished to talk about my salary. To my astonishment, his answer was that he did not care to discuss the matter. I was hurt. I told him that the matter of my salary was of mutual concern and that I had as much right to request discussion of the matter as he. Further, I gave him one week's notice and told him about the offer that I had received six weeks earlier. The following Monday, Szabolcsi offered me $25.00 per week. I turned him down.

I had decided to enter business on my own. As a result of some correspondence, I had learned about the best markets for Hungarian goods. If, in Szabolcsi's business, we received many orders from a certain geographic section, it was proof that the section was not visited by Hungarian peddlers. I had decided to become a peddler and visit those places.

Szabolcsi repeated his offer every day and on Friday, he offered me $40.00 a week if I stayed with him. This was very tempting. It was an unheard-of salary, and I would have been the best-paid man in Hungarian circles. But somehow, it would have been against the grain to remain.

I patiently told Szabolcsi that I never would work for a man who considered my services worth $40.00 a week but was unwilling to pay even $25.00 voluntarily when he knew that I was offered as much elsewhere. Much later, Szabolcsi told me that his refusal to increase my salary was based on the information that he received from the branch manager of the *Szabadság*; the info was that I was honest but stupid, and Szabolcsi took it for granted that I would be afraid to change jobs. This after, in my "stupidity," I had assisted him in his literary activities.

When I turned down Szabolcsi's last offer, he asked what I did propose to do and pointed out that competition was very keen both in the banking and mail order businesses. He was somewhat embarrassed when I told him that I would try something entirely new, something that no one else had tried amongst Hungarians. I would try being—honest. Honest to the extent that I never would represent 10 karat gold as 14 karat or bone on a prayer book as ivory.

CHAPTER 11

On the Road Again

★

I BOUGHT A SUITCASE and also had one made to carry a small gramophone and six records. I selected a variety of stock, partly in Szabolcsi's store and partly in the novelty houses on Canal Street. To pacify me and to secure my trade, Szabolcsi put an ad stating that I represented his firm in the Hungarian newspapers. Hungarians may trust me. The ad was a great help.

Carrying the gramophone and the records in one hand, the prayer books and other imported articles in the other, I took the train for the nearest promising colony, Flemington, New Jersey. Knocking on the door of the first Hungarian home was very hard, but I went in, unpacked my gramophone, played my records, and within a few minutes, several of the neighbors joined us. I sold the gramophone and the records, some prayers books, and sundry other articles, and I left the house with orders for two gramophones for two neighbors.

I made over $15.00 above my expenses, and the orders meant a profit of close to $30.00. Within four days, there was a gramophone in every one of the nine Hungarian homes of Flemington. And within a week's time, I doubled my original $140.00.

But the work was hard. It was no cinch to carry a gramophone and six records in one hand and balance it up with another load of books, etc., in the other hand as I walked the three miles from the train station to the Hungarian colony. But hiring a taxi was an un-dreamed-of luxury that never would have occurred to me, a peddler.

I invited one of my buddies from the Gleason Mine who also was a veteran of the *SS Carpathia*, John Kormendy, to join me in my peddling venture. Since John was broke, I promised to split the profit even until he had enough money to invest half of our capital. From then on, we carried four suitcases and a much larger stock.

There were few isolated places undiscovered by Hungarian competitors close to New York, and with my policy of absolute honesty, I could not meet the competition of numerous other peddlers. I decided, therefore, to go to the mining places of Virginia and West Virginia which were the source of most of the mail orders to Szabolcsi.

My decision to go to the Virginias was a daring venture. Train fare to Huntington, West Virginia, was about $15.00 each for John and me, but from there on, one mining place was only about twenty miles from the other. We did very well amongst the miners. While Hungarians in New York were sick and tired of peddlers following on each other's heels, the miners in Virginia and West Virginia were very appreciative. They also had much more money than factory workers. The peddling was a killing job for mines usually were located on a spur three or four miles from the main railroad line. John and I had to walk long distances with our heavy loads, but we were young.

The southern coalfield was unorganized, and operators were scared of strangers. Each operation had a few mine guards employed, and all of the guards were furnished by the same Baldwin-Felts Agency of Bluefield. Since no attempts had been made by the union to organize the southern miners, the guards had to find straw men to hang on to their jobs. At times, John and I were chased out of a camp by those guards who no doubt reported that two dangerous organizers were waylaid by their alertness. On smaller camps, John and I were not welcome because we offered, no matter how small, competition to the company store. The policy was that miners should spend all of their earnings in the store of their employer.

Necessarily, we had to conduct the business after quitting time and were forced, more often than not, to spend the night on the camp, usually in a hospitable Hungarian boarding house. I found the boarding houses terrible after my rather modest home at the Varhelyis, and if there were one possible chance to walk to the nearest little mining town, I availed myself of a decent room. John Kormendy left me after two trips to the Virginias. He told me frankly that he could not take it, although his savings of more than $150.00 likely had something to do with his decision.

I traveled alone from then on, re-visiting customers and looking for new places. I often was disappointed by finding only one or two families on a camp, and each camp meant at least a day of finding unexpected competition. I also tried the mining camps of Pennsylvania, only to find the camps invaded by peddlers from New York as well as from Pittsburgh. Peddling was a discredited occupation in Pennsylvania; every peddler was received with suspicion. If I were requested to show a watch to a Pennsylvania miner, someone in the house would question if the watch kept correct time. If I offered a meerschaum pipe, someone would certainly speak and say the pipe looked imitation to him or her. I was offended by such remarks and was not used to hearing them in the Virginias. Nor could I stand the bargaining habit that Hungarians were used to in the Old Country. Most of the peddlers were willing to put up with insults and were also in the habit of doubling the asking price, knowing well that they would have to come down. My one-price policy wasn't effective in Pennsylvania.

I confined my activities to the Virginias and the Hocking Coal Field of Ohio,[2] with an occasional trip to Zanesville, Dayton, and Newark, Ohio. As I became acquainted in these areas, I started a revolutionary policy. Since most of the immigrants mailed their money to the Old Country after pay day, it was hard to sell them anything amounting to more than a few cents between pay days. The coal mines paid but once a month. Therefore, as soon as I became acquainted with the boarding bosses on my territory, I offered all of my items on credit and left a return envelope behind. My customers were astonished at the trust I placed in them. Up until then, they had sent money in advance with the mail orders or paid cash to peddlers, and my new system was highly appreciated. I think that some purchased a razor or a prayer book for the sole purpose of trying me out to see whether I really meant what I said.

Confidence begets confidence. I received orders for things I never dreamed to sell — Hungarian paprika, Hungarian jelly, seeds, rose bushes, shoes, and all kinds of other items. When I returned to New York, I mailed every order without any loss to myself.

It was not necessary for me to run into New York every day to replenish my stock if I sold a certain prayer book or some other article. My customers were not afraid to place their orders and if an imported article were ordered, the order simply was forwarded to Szabolcsi and the order was fulfilled. Not having anything else to do during the day except to go to the next mining camp, I tried to sell goods wholesale to company stores. I requested some of my Hungarian friends to go into the various company stores and ask for long-stem Hungarian pipes or a Hungarian meerschaum pipe. I assured my friends that they were safe to do this because the company store did not carry the articles for which my friends asked. When I made my next trip to the store, I offered the asked-for goods to the buyer, and he remembered that some customers previously had asked for such goods. Then, I sold a small lot of requested goods. I was not afraid to destroy my retail trade because for one reason, I charged a stiff wholesale price. For another reason, I was sure of the loyalty of my clients. I did not have much sympathy for company stores.

I did big business with leeches used by immigrants for a little blood letting. I usually sold leeches in small-town drug stores by using the same tactics as in the company stores. Once, I struck an unexpected rich vein in this gold-bricking wholesale business. When offering my small smoker's articles to the buyer in the Norton, Virginia, company store, I had to open my suitcase. The buyer then saw a big prayer book decorated with "ivory." The buyer said that he had seen several such books in the hands of foreigners and wanted to know whether all foreigners could use them. Hoping that I could unload about six prayer books on him, I assured him that a "hunk is a hunk," and all could use the same kind of book. He ordered one hundred of the prayer books.

There were Catholic and Protestant Hungarians and though the Protestants were not divided into so many denominations as in America, there were some Lutherans and Baptists and each denomination had several prayer books. I had to carry about twenty different kinds of prayer books which were about only one-third of the books generally used. And this Norton, Virginia, man ordered one hundred prayer books for Catholics. The order meant a net profit of $100.00 for me, and I was elated. But, since I knew that there were only four Hungarian families in the nearby Dorchester mine, and since the company store was smack opposite the railroad depot, I was afraid to go to Norton again. To my great relief, I heard that the company store burned down about eight weeks later.

I made good money as a peddler but could not save much. I spent a few days resting in New York between peddling trips, and I spent much money exploring new territories. I wanted to see as much of the United States as possible. When a woman told me about winter life and the six feet of snow in the Upper Peninsula of Michigan, I immediately took the train to give the Upper Peninsula the once-over. Such trips were expensive, but sometimes they had interesting results. When I was traveling north at the end of December, I stopped over for Christmas in Detroit. I engaged a room in the Young Men's Christian Association (YMCA) and was deeply touched when strangers were treated not only to a fine Christmas dinner, but also to a theatre ticket to see Otis Skinner in *Kismet*.³

I found the Szabo family with whom I boarded in Tompkins Cove, New York, at one of the copper mines near Calumet, Michigan. Of course, I established a land office peddling business there. I worked at Calumet for about six weeks, did more ice skating than work, and I got acquainted with some Indians—also Indian girls—for the first time.

My greatest ambition since arriving in America was to become an American citizen, and I was granted the privilege of citizenship in April, 1913. The same summer, when I had about $600.00 and some stock, I decided that it was about time to visit Hungary and to see Mother and—Icza. I wrote to my mother and said that I would be home on August 15, the homecoming day of Mátraverebély, and I conveyed to Icza the same intelligence. I received a letter from Icza's mother within four weeks. She told me that her daughter married an army officer two years earlier and that my letter was not forwarded to her. Icza's mother said that I would be welcome to visit Icza's family, but I should not approach Icza. My dream of close to eight years was over, but I wanted to go home just the same.

I stocked up on my peddling items, invested all my money in watches, shawls, etc., and started out to make a last grand slam before my trip to Mátraverebély, the cost of which I estimated to be $1,000.00. I took the train for Hocking Valley to start with my old friends, the coal miners, but work was slack and business disappointing. I was impatient to work from mine to mine, and I visited a few families each day. I decided, therefore, to confine my activities on this trip to the large Hungarian colony of Dayton, with a stop-over at Zanesville, Ohio.

I arrived at both cities before pay day, but I did not mind. Selling everything at less than the usual prices, but with still a respectable profit, enabled me to get rid of most of my stock within a few days. I took the train eagerly for New York and left Ohio with more than $1,000.00 in outstanding accounts to be received by mail after pay day.

I went around to shop for steamship tickets when I returned to New York, and I happily was anticipating my trip. I also did some window shopping as I thought of the pending arrival of plenty of cash. Three days later, I awoke and read large newspapers' headlines that announced the famous Dayton Flood of 1913.[4] Included in the reports were details about the destruction of the famous Y Bridge of Zanesville[5] and the vivid descriptions of thousands of homeless people. Hungarian districts in both cities were destroyed, and my customers now were taken care of by the Red Cross. I was ruined on the eve of my victorious visit to my folks! If I were downhearted and bitter then, it did not take me long to recover.

When attending the night school in New York, I learned that ". . . the man worthwhile is one who will smile,/When everything goes dead wrong."[6] "Smile, darn you, smile, or perhaps you may go as far as laughing," I told myself. And within two hours, I was in Szabolcsi's store buying a new line of stock on credit. One has a lot of self-confidence and energy when one is twenty-five years old.

Not having enough cash to go down to the Virginias, I went to Roseton, New York, a tremendous brick yard with hundreds of Hungarians who did man-killing work. But they made good money, about $3.00 daily, which was almost as much as coal miners earn. However, Roseton was a favorite spot of all the Hungarians who were peddlers. Roseton was run over with peddlers as if the locusts had visited the place, and competition was very keen.

Since I had to buy everything from Szabolcsi, instead of novelties at Canal Street, I was at a tremendous disadvantage. Feeling my desperation, I approached Negro workers with my watches and razors. To my great and pleasant surprise, the Negro workers were willing buyers. I had a gold-rolled watch with me that did not sell at Dayton for Hungarians went in for silver watches. But the gold watch took the fancy of one of the Negro boys. The watch cost me $1.40 but, expecting bargaining, I set the price at $10.00. Without blinking an eye, the Negro boy offered me $2.00 down and $2.00 on each pay day. I knew darned well that it was July and there were only two more pay days in the year because the plant was closing in September. But I did not care, and the sale was made to mutual satisfaction. I made $0.50, the boy had a $10.00 watch for $2.00, and within minutes, I received orders from his friends for twenty-three additional watches.

Two days later, all the watches were delivered on the same terms and netted me over a $10.00 profit, besides the expenses. I also sold cosmetics for over $30.00 at $0.40 for each article purchased by me in the dime store.

Believe it or not, I had to return on the same day to nearby Newburgh to load up once more with cosmetics from the dime store. When I finally left Roseton, I had enough to pay my bill at Szabolcsi's plus a $200.00 profit with most of my original stock from Szabolcsi intact.

While shopping the next day on Canal Street, I passed a pushcart loaded with picture postcards which were beautifully decorated with kissing pigeons and various other symbols of love. The postcards were all the rage about fifteen years earlier. I purchased the whole load, about six thousand cards, for $12.00, and I purchased one hundred pairs of eye glasses at $0.10 per pair from another push cart. I loaded up the postcards and sundry other novelties and left for the dense Hungarian section between Perth Amboy and New Brunswick, New Jersey. At first, I did not bother with the retail trade. I visited small Hungarian, Slovak, and Polish merchants and offered them the cards at $3.00 per hundred; I pointed out that each card was worth $0.10 — $0.15 in retail. Since the cards did sell about fifteen years ago at such prices, I got rid of my load in three days. Then, I started out with my eye glasses. I went from house to house and offered the privilege of trying as many pairs as the prospective purchaser wished in order to select the pair best fitted to the purchaser's needs. I sold about eighty eye glasses at the price of a lone dollar within two days. The rest of the eyeglasses were sold to a Hungarian local merchant for $0.35.

I returned to New York with $400.00. This was not much, considering the fact that I had to buy some clothing and that the cost of the round trip was close to $200.00. But my time was running short, and I could not think of disappointing my mother.

One of my friends, a Mr. Gyulay, promised to send me enough money for the return trip to America if and when I decided to return from Mátraverebély. Thus reassured of feasibility of the trip, I sent to a shipping agent who told me that a Dutch steamer would leave the next day. One of my friends, an old peddler named Mr. Lajtos, requested the privilege of traveling with me. We purchased two third-class tickets for the steamer *Ryndam*.[7] Each ticket cost $41.00. Of course, in those happy days, passports, visas, and other stupid formalities were not necessary. But, I took my certificate of citizenship, and Lajtos took his first citizenship paper which he secured a few hours before we embarked for our journey.

The minute we went on board, the smell of the boat made me seasick. Lajtos and I then exchanged our tickets at the Purser's Office for second-class tickets by paying an additional $20.00 on each ticket. The second-class accommodations and food were far beyond everything previously enjoyed by me in my life. The sea was calm, as it is mostly calm in the summer, and Lajtos and I had a wonderful eight days.

There was a good-looking Hungarian girl in the second class, and Lajtos and I spent many pleasant hours in her company. But there was also an old Hungarian peasant woman, whom I disliked from the start, and she managed to create an unpleasant incident at the end of the trip. The old lady had four sons in Trenton, New Jersey, and instead of each son's going home to visit his mother, all of them financed her trip to Trenton for a six months' visit. The old lady was on her way back home to Hungary. She was dissatisfied with each of her daughters-in-law, and according to her own story, she managed to "open the eyes of her sons," meaning that she created endless troubles between men and wives. She was bragging about her accomplishments.

During the best part of our trip, when the steamer was majestically floating up the River Maas and nearing Rotterdam, the old lady requested the privilege of staying with us overnight in Rotterdam. I could not tell her to go to the devil. When she was told that the fare would be a Florin (the equivalent of $0.40) for each of us, she started to holler while we were traveling on the carriage. Lajtos roughly told her that he would kick her off unless she became quiet. She was afraid to stay in a room alone in the Rotterdam hotel and insisted to share my and Lajtos' room. Finally, a connecting room was given to her.

But the real trouble broke the next morning. Leaving the old lady and Lajtos at the depot, I went out to buy some bread and cold cuts for the twenty-four hour trip. At my station of life, I would not dream of taking meals in the dining cars. While I was out, she had Lajtos buy her a ticket, and on my return, I purchased two tickets for Lajtos and myself. When I told Lajtos to hand over 25 Florin for his ticket, the woman started to holler again. She dumped herself on the floor, cried that she was robbed, and pulled a fit similar to those of the epileptics. We soon were surrounded by a large number of people, and I was helpless. I did not know what was wrong with her. Finally, and luckily, there was a linguist railroad man at the station who, hearing the Hungarian language, came to see what was going on. Between crying and cussing, the old woman told him that while the price of the ticket to Budapest was 25 Florin, Lajtos had her pay him 40 Florin.

We finally discovered that the old woman had tickets for two. Lajtos could not speak English, and while he was buying her ticket, she was dancing around him at the ticket window. The ticket agent therefore took it for granted that Lajtos and the old woman traveled together and needed two tickets.

This unpleasant matter was adjusted before Lajtos and I were lynched by the threatening audience or arrested by the law, and the ticket price difference was refunded to the old woman. But since then, I have tried to avoid Hungarians when I travel. I do not wish to subject myself to similar unpleasant experiences.

We left Rotterdam by a fast train, traveling in third class, of course, and twenty hours later we passed Vienna. The countryside was very nice, especially the picturesque windmills in Holland. But even Germany looked good then.

At the Austrian border, we had some difficulties. Lajtos had loaded up with American cigarettes and outrageous duty was demanded by the custom officer. When Lajtos got mad and destroyed his cigarettes, the officer tried to fine him for doing so.

Lajtos and I reached the Hungarian border about 2 P.M. on a Sunday. Who can blame me for having tears in my eyes when I first saw the Hungarian types on the cars of the Royal Hungarian Railway, or when the conductor asked for our tickets in Hungarian? Noting that our tickets were from Rotterdam, the conductor knew that we were coming from America. Upon receiving two American cigarettes from us, he closed the door of our compartment and put up a sign that said our compartment was a "Service Compartment." The sign meant that our compartment was reserved for the two of us.

As we neared Budapest toward evening, the corridor was chock-full with returning excursionists. I was embarrassed to occupy eight seats with Lajtos. When the conductor passed, I motioned to him to open the door. When he did so, I told him to remove the sign from the window. I saw the contempt on his face for he must have realized that we were no gentlemen if we were unwilling to use the privilege he bestowed on us. I soon regretted my impulse to have the sign removed for Lajtos immediately started to talk to the other passengers, and I had good reason to resent his endless chatter. Like so many returning Hungarians, he was not satisfied to relate the wonderful American life as it was. He was telling tales that raised the hair on my head. And in Hungary, like in most Eastern European countries, there was a law against "tempting to emigrate."

We finally reached Budapest at about 8 P.M. I will not attempt to describe my feelings; it can't be done. I had my brother Jakab, the cabinet maker, in Budapest, but I did not wish to go to him at once. Not knowing his circumstances, nor his wife, I did not wish to embarrass them by dropping in on them late in the evening. We went to a hotel instead, a hotel right at the Western depot. I also remembered the bitter opposition Jakab felt before I left and his refusal to take the same coach with me from Mátraverebély. I did not know how I now would be received by Jakab.

Lajtos and I were tired after a trip that lasted over thirty hours, instead of the twenty-four that we were told the trip would last. Following a goulash supper — with gypsy music — we retired.

When a knock on the door awakened me the next morning, I realized that I never could live in Europe under any circumstances. An old woman came in for our shoes. I never had been in an European hotel, and I was not familiar with the custom of leaving the shoes in front of the door to be cleaned. When the poor woman came in, she greeted us with "I kiss your hand esteemed lords," and my stomach turned upside down listening to the servile greeting. The whole unspeakable Hungarian caste system came back to me, as did the impossibility of crossing the line from one caste to the other. If I had returned to Hungary with the intention to stay there, that one greeting from the poor woman would have changed any such intentions.

Prior to breakfast, I ordered a florist to send a bouquet to my unknown sister-in-law. Breakfast then furnished another unexpectedly unpleasant incident for Lajtos and me. Neither rolls nor bread is included in the price of any menu in Hungary. When we attempted to pay for our breakfasts, we neglected to state the number of rolls consumed.

"And the rolls?", the headwaiter asked suspiciously.

Lajtos did not need any other excuse to start going. He told the headwaiter that we came from America and that we were not accustomed to paying separately for rolls. Thus, we were negligent in stating the number of rolls consumed. But Lajtos did not stop there. He also told the headwaiter that we arrived only last night, that we were on the sea for eight days, etc., etc., and before Lajtos was finished with the first sentence, there were four other waiters listening to the tales. Returning Americans were rare in those days, America was much farther away than now, and innumerable questions were asked by the waiters. I had to be rude to put a stop to the Lajtos stories. Bidding him good-bye, I went to my brother.

My brother was advised of my presence by the bouquet, and the greeting was friendly, but not warm. I was on trial in his eyes. My cheap new suit had become shabby, and I told my brother that, first of all, I wished to order a new suit of clothes. I requested the address of his tailor. I was then reminded once more that people with money were lords in Hungary for an apprentice was sent to the tailor ordering him to appear at once to take measure for a suit of clothes.

My brother was well-to-do, even wealthy by Hungarian standards. He had forty journeymen working for him, and his wife and kid were away in a summer resort. Thus, sending the bouquet was a futile and wasted gesture.

Following a short visit with my brother, I went to the American Consul to ascertain myself as an American citizen so that they could not grab me for the Austro-Hungarian Army. The Consul assured me that I was safe but advised me to apply for a passport and avoid any possible complications. I paid $2.00, and the passport was mailed to me within five days from the American Embassy of Vienna.

My brother and I went shopping after lunch to buy presents for my mother and for numerous offspring of sisters and brothers. While going around Budapest, I realized that Budapest had become a really beautiful city in my absence. During the Balkan War,[3] Hungary was exporting food and war material at exorbitant prices to both the Turks and to the Allied Balkan Nations. Everyone was prosperous. Perhaps it was the only short period that Hungary was prosperous during the twentieth century.

My brother was trying me out throughout my two and one-half day stay in Budapest. He tried his best to catch me in misstatements. One evening, he took me to a relative where "just by chance" an English-speaking guest also was present. The guest was requested to talk to me in English, "just to listen to the sound of it." Well, my English was poor, but the guest's was worse, and I came through the test with flying colors.

Next day, August 13, 1913, I took the train—in my new suit—for Mátraverebély.

CHAPTER 12

With Mother

★

MOTHER WAS OLD, well over sixty years old, and I planned to arrive before August 15 in order to spare her the excitement of the last few hours. On the other hand, I was afraid to shock her with an early unexpected arrival and thus wired to Pásztó to have my brother advise her by messenger that I would arrive at 1:00 P.M. Mother soon got wise to my arrival time tricks so when I made future trips home, I usually arrived several days earlier then I was expected.

The four-hour trip from Budapest to Pásztó seemed like four years. It was far longer than the voyage from New York. But finally, the train pulled into Pásztó. Two of my married sisters and their husbands and my brother and his wife, as well as several of their children, joined me there, and I had to get re-acquainted with the young men and young ladies who were children when I left.

The last twenty-minute ride from Pásztó to Mátraverebély was the most exciting. I traveled the same route innumerable times when I was attending school, and I knew and recognized every tree and every stone along the way. I resented being bothered by the questions of my relatives who did not leave me alone for my meditations.

Have you ever been away for six years and then returned to your aged mother? If so, there is no need to describe the tearful meeting. If not, description would be a useless attempt. When I reached home, my mother took off my glasses for she could not believe that a young man of my age really could need glasses. Then she started to scold me for fooling her about my arrival. She was whitewashing the walls in my honor, and I found her in the midst of this work. But I knew that she was pleased to have me there, early or late.

Dinner was served and after dinner, the presents were distributed. The presents included a long gold chain for Mother and watches for the youngsters. Mothers are funny sometimes. My mother suddenly burst into tears and bemoaned the fact that Father was not alive to see the return of the prodigal son, the ex-black sheep of the family.

The Pásztó folks had to leave at 4:00 P.M., and I then unpacked the other, the real present for my mother. The real present was the one poppy seed biscuit that I had saved for close to six years, the biscuit into which the tears of my mother were dropped when she prepared the biscuits for my first journey. When I was homeless, Mr. Wechsler, the Hungarian dry goods store keeper of Avenue A, kept the poppy seed biscuit for me. I told my mother that her biscuit brought me back, as she said it would, and, of course, both of us had a good cry at the sight of the stone-like, dark biscuit. The biscuit was shown to me by my mother whenever I went home on succeeding visits.

As the afternoon went by, I told my tale. I spoke frankly, omitted nothing, did not deny that I was a hobo at one time, and did not hide the fact that I was all but broke. However, I assured my mother that I was healthy and young, knew my way around America, and could make a living anytime, at anything, on my return to America.

While my sisters and brothers were not elated by the fact that I was a peddler—an occupation below even that of the peasant in Hungary—and could not see for the lives of themselves my preference of a peddling life against working in an office in Hungary, my mother was pleased. She was happy that I did not put on airs and that I was frank, and nothing else mattered with her. For that matter, what mother would not be pleased with the first visit of her son after an absence of six years?

My family could not conceive, much less believe, that I could not have been any prouder if I were the prime minister of a country. My family also could not conceive that I did not, and still do not, know or recognize any difference between occupations or, as we say, in one's station of life, except the difference between honest work and a dishonest way of making one's living. I have close friends amongst "captains of industries," but none closer to me than my coal mining friends. And I am still very proud to have the friendship of a peg-legged street sweeper in Detroit who lost his leg in a coal mine.

As I traveled from the train station to Mátraverebély, I had a vague feeling that I would be disappointed. I noted with a sinking heart that everything looked small and dirty. I suppose everyone has the same feeling on a return to the scene of his childhood after several years of absence. Toward evening, the neighbors, friends, and schoolmates came in bunches. I was amazed that I recognized my schoolmates solely because most of them looked as their fathers did when we were children.

I had a tough time side-stepping a good many questions while keeping in mind the strict Hungarian law about "tempting to emigrate." I had to tell my visitors that the poor are poor anywhere, even in America, and laborers also have to work hard in America. Meantime, my heart was bleeding for my family and friends because I knew how much better off they would be in our America.

Instead of telling my visitors the amount of money that one earned with honest work in America. I told them my experiences with carrying hitz. Instead of telling them that laborers eat meat three times a day, I told them about the unattractive life in the boarding houses and the hardships in being away from one's loved ones. I would have liked to cry out, to tell them to come join me in America and escape the serfdom to Count Almassy, but I had to be careful.

Next morning, I called on the village notary and sure enough, he reminded me of the much cited "tempting to emigrate" law. Therefore, I was even more careful lest I should say something nice about America. Interestingly, on the following day, two gendarmes[1] who were in Mátraverebély "by chance" called on me. They engaged me in a friendly conversation about America.

One of my former schoolmates suddenly was stricken with tuberculosis while I was at home, and on my fourth day home, I started to work in his place at the thrashing machine. I carried the heavy sacks full of grains to the second floor of the warehouse without difficulties for three weeks and thus increased my popularity in my village, which was not my intention at all. Later, I visited my sisters and brothers in Szombathely[2] and Debrecen,[3] and all of them came to Mátraverebély several times while I was home. I would have had a glorious time, but there were two flies in my ointment. First, I was constantly besieged and battered to change my plans and remain home. When I pointed out my sad experience before going to America, and the endless weeks spent in looking for a position, money was offered to me to establish myself in some business. Two of my brothers had become well-to-do, and, for Mother's sake, they were willing to loan me enough money and set me up in a business. Perhaps they were willing to believe that I had changed and ceased to be the black sheep that I was in my younger days. But remembering my unhappy years while indebted, when the bitter days all but wiped the smiles off my face for life, I shuddered even at the thought of accepting loans. Only my mother kept silent. Perhaps she was hopeful that I might remain in Mátraverebély, but yet she approved of my reluctance to borrow money.

The second fly in my ointment was the issue of marriage. I constantly was pestered by offers of suitable marriages when I was in Mátraverebély. My family did not know anything about my love for Icza whom I did not see until much later, in 1946, and my family kept on mentioning girls after girls with respectable dowries. My family was quite surprised by my contempt about marrying for or even with dowries, and they also were surprised by the fact that, in America, the dowry custom was unknown. I was not then familiar with the lives of multi-millionaires amongst whom the custom prevails even in America. My family explained to me that the dowry is a just requirement to offset the cost of raising boys and educating them or teaching them a trade, and that I therefore could select a girl with enough money to establish me in some business in grand style.

While I did not consider marriage for a minute, for the Icza wound was fresh, there were times when I was swayed by the constant attacks to remain home, especially when thinking of my aged mother. I once thought that even if I remained in Mátraverebély, I could return to America after Mother's passing.

One day, I traveled from Mátraverebély to Budapest and upon my arrival, my shoes were muddy. There were two men shining shoes for the public in Budapest, one man at each of the railroad stations. I had to stand while the shoe-shining man sat in his small box and cleaned my shoes. I then thought of the idea that perhaps an American style shoe shining parlor would go over well in Budapest.

I later went to my brother, Jakab, and I told him about my shoe shine experience in Budapest. I then described to him the American shoe shining parlors that contain mirrors and comfortable chairs where the customers sit while the standing shoe shiners clean the customers' shoes. I remarked to my brother that perhaps I ought to stay home for a while and show Budapest a little of the American customs.

My brother was enthusiastic about my plan. He said that while poor people shine their own shoes, and the well-to-do have servants, the people of Budapest like to put on the ritz and would undoubtedly take to the idea of having their shoes shined in an elegant parlor. He started to draw construction plans, for his business was to build portals for stores, and he said that I could hire a few cheap boys to do the work. When I said that at least at the beginning of my business I would shine shoes myself instead of hiring boys, my brother grew rigid and red in the face. He said that if I were to do that, he would be compelled to move out of Budapest.

I was mad clear through!

The social castes in Hungary at the time were: the aristocrats, the common but wealthy land owners, the army officers who usually were recruited from the two top castes, the lesser gentry who had but a few hundred acres of land, the men in state, county, and city services, the professionals, the office workers in private employments, the merchants, the artisans having their own shops (the caste to which my brother belonged), the peasants having a very few acres of land called midget farms, the factory workers, the land-less peasants, and the peasants hired by the year as servants of the landlords. My brother belonged to the ninth caste from the top and felt that my shining shoes would force him out of Budapest because he could not live down his brother's doing honest work in the same city.

I blew up and very impolitely told my brother in an unprintable four-letter word what he could do with Hungary. I wired the same day to my friend Gyulay for the promised loan for $60.00. I also accepted a loan of 200 Kronen, $40.00, from one of my sisters in Szombathely who was in a fairly comfortable financial situation then.

One day, Lajtos came to visit my folks and me in Mátraverebély, and it was the first occasion that I was criticized by my mother. Not realizing that I was frank with my family, Lajtos started to bluff and told some lies and half-truths. When Mother and I were alone, Mother told me that she was very happy because I was truthful with her. But she couldn't see how I could select such friends as Lajtos.

To cut Lajtos' visit short, I told him that I would like to visit his village. The next day, Lajtos and I went together to his village. While visiting

Lajtos' village, I realized some of the reasons that Hungarians returning from America are very popular in the Old County, and I also discovered the sources of the exaggerated ideas some Europeans had about America. When Lajtos appeared in his village where he had only distant relatives and friends, he told them that he would leave them his fortune. He did not mention that his fortune consisted of a very few dollars and a return ticket to America. I spent exactly eight hours in Lajtos' village.

I became acquainted with several newspapermen and some government officials during my stay in Budapest and, off the record, I told them a thing or two about America. I never failed to show each of them a letter which I received from America's State Department when I inquired about my status in Hungary regarding military services. I pointed out the salutation of "your obedient servant" before the signature.

This then customary salutation on every letter sent by any of America's State Departments amazed the Hungarian newspapermen and government officials for the salutation was proof that public servants really feel themselves to be just that in America, very much unlike in Hungary. If a Hungarian petitioned any of the Hungarian ministries, he or she never wrote a letter, only a petition, to the minister, and the size of the petition was officially prescribed as well as the method of folding same. The outside of the petition had to state that it was addressed to "His Excellency," the minister of such and such, and contain the petition of "your humble servant" such and such. I am sorry that the custom of the old American State Department salutation was done away with during the Coolidge Regime. Signing their names under "your obedient servant" must have had some salutary effect upon America's secretaries.

While I was visiting Hungary, the Governor of the state of New York (Governor William Sulzer, if I remember right), was impeached. I did not fail to call to the attention of my newspaper friends that removal of a public official was possible in a democracy.

But everything comes to an end. It was October, the leaves were falling, the fields were white in the early mornings, and I decided to leave Hungary. Upon receipt of the loan from America, I purchased a ticket for a German steamer, the *Pretoria*,[4] because she carried only two classes. I was still enough of an European to resent having people in a class above me on the same steamer.

All of my sisters, brothers, in-laws, and their offspring gathered in Mátraverebély once more on the day of my departure. My mother baked the poppy seed biscuits once more, crying and dropping her tears into the dough just as she did when I first left her.

My train left at 4:00 P.M., and about noon, many of my schoolmates and friends came to say "good-bye." Every girl and young woman presented me with a lovely embroidered towel made of homespun linen. The towels are still the most cherished decorations of my home.

Fifty people came to watch my train departure, and my crying mother again whispered that she never would see me again. I was relieved and happy when the train pulled out. If possible, I loved and admired my mother even more than I ever had in the past. It takes a brave and unselfish mother to send her son off with the belief that the departure will be forever because she believes that her son will fare better in a far-away country.

Lajtos was on the same train. He, as well as several other passengers, were very helpful in disposing of the mountains of roast chicken and cakes given to me by my family, but none of them were permitted to taste the poppy seed biscuits into which my mother's tears were mixed. Gendarmes went through the train at the Hungarian border to look for immigrants without passports. They saluted me when I presented my passport. They likewise saluted at the sight of Lajtos' first paper with the English text and the American eagle. How could the Hungarian gendarmes know the difference between the two strange documents?

We stopped in Berlin for a half day, and I did not much like the town. Nor was I very much impressed by Hamburg, except by the proud sign on the building of the Hamburg-American Line[5] stating, "mein feld ist die welt" or "my field is the world."

The *Pretoria* was supposed to make New York in eight days, but we encountered a heavy storm and a screw or something broke. So, the steamer limped into port on the nineteenth day.

However, we did have a jolly time on the voyage with the exception of a seasick mess which lasted about eight days. There were ten other Hungarian passengers, most of them young. I got acquainted with a young boy about seventeen years old who told me that he was somewhat scared for he did not know anyone in America. He was a pleasant chap, and I assured him that I would look out for him. Indeed, a few days after our landing, I secured him work. Shortly after, he started to attend art school. Years later, he was commissioned to decorate the Philadelphia Museum of Art. Still later, he became a sculptor in Paris. Now, Mr. Fulop[6] divides his time between his Paris and Hollywood studios.

Mr. Fulop's sister, Ilona Fulop,[7] who later became a well-known Hungarian newspaperwoman, soon joined her brother in America. Ilona also was a writer, and her *Spring Shower*[8] became a successful American film. Two more Fulop brothers came to America and opened a pottery studio in Van Nuys, California. All of the Fulops remain my close friends to this day.

As a result of my absence, Mr. Varhelyi got himself a partner for his shoe store. I moved away and rented a two-room tenement apartment for $11.00 and furnished the same at a cost of about another $11.00.

Taking my two suitcases in hand once more, I resumed my old peddler trade. However, I had the queerest experience on my first trip back on the road. Another peddler who sold patent medicines and I went to Proctor, Vermont, where some Hungarians were employed in a marble quarry. When we visited the first Hungarian home in Proctor, we found a group of weeping folks and a coffin in which the woman of the house was laid out. Of course, we did not offer our goods, but we did have to tell the residents of the home the object of our presence on the place. When I mentioned prayer books, one of the women, an elderly lady, had a bright idea.

The lady told me that she and her family were sick with worry because they did not have a Hungarian minister. Immigrants would have been afraid to approach an American minister in those days. The lady then proposed that I conduct the funeral. It was of no use to explain to her that I could not conduct the funeral and that it was completely outside of the scope of my activities. She would not budge. She pointed out that by selling prayer books I must be familiar with their contents, and, being a learned man, it would not be too difficult for me to lead the prayers.

When the widowed husband joined the lady's begging, I could not refuse. Luckily, they were Protestants, and all that I had to do was select prayers and songs and compose a last statement in the name of the departed. The last statement said the departed woman's good-bye to husband, children, near and far relatives, and friends and neighbors. This last statement was a custom in the Old Country. The last farewell, said in the name of the dead woman by me, and my special speech were huge successes, and the whole audience cried. I would have been a very poor individual if I had failed to make them cry when I referred to the Old Country and to the grave in the strange land. I refused the $20.00 offered by the widowed husband for my services, but I sold enough goods to be amply compensated. Carried somewhat away by my success at the funeral, I almost decided three weeks later to become a minister.

My next peddling trip took me to Scranton, Pennsylvania, and I met some strange Hungarians at a small place near Scranton called Brinza. I was invited to dinner at Brinza, and I was surprised to see that boarders were drinking water instead of whiskey. Hungarians, as well as other immigrants, were heavy drinkers. Each boarder in each boarding house had his own $.90-$1.00 jug of whiskey next to his bed, and many boarders consumed several jugs a week. Each jug held a gallon of whiskey. I then was young and intolerant, and I did not stop to think that all that the immigrants had for amusement, recreation, and—forgetfulness—was whiskey.

The immigrant men worked very hard. They constantly worried about their families, wives, and sweethearts who were left behind at home. They especially worried when one of them was advised that he would find a new child in his home upon his return, a child fathered by a man who had remained home. When one immigrant in a camp was advised of such a happy event, everyone started to worry, and several men rushed home at once. Ex-GI's know how pleasant it is to leave a young wife behind for several years. The ex-GI's can measure some of the prices immigrants had to pay for the privilege of contributing their labor to building up American industries.

But as I said, I was young and intolerant, and I worried much about Hungarians' spending so much of their hard-earned money on alcohol and thus undermining their health. The water-drinking Hungarians at Brinza told me that they were newly-converted Baptists and that there was a Hungarian Baptist seminary on the camp. The existence of this seminary was almost unknown by other Hungarians in America, and it certainly was unknown to me. I went to the seminary and found eight very earnest young men, one of whom I had met the previous summer at the Roseton brickyard where he was working as a laborer.

The Baptist faith was not an accepted religion in Hungary. While accepted religions were supported by the state, tolerated denominations failed to receive any state assistance. There were very few Baptists in Hungary. Some Hungarians seeking solace, some seeking spiritual food in their hard lot, and some who had been away from churches of their own faith and also from the social control of their villages and from their folds, turned to other denominations. The new Hungarian Baptist converts at Brinza took their religion very seriously, and their zeal was at a very high pitch.

Since Hungary was unable then to furnish Baptist ministers, the country recruited new converts and induced them to become ministers. Later, the

Baptist religion spread in Hungary, and a few very able ministers came from there after the First World War. But at the time of my visit to the seminary, the students were uneducated peasant boys, and the curriculum was composed of that of a four-grade high school, besides the religious aspects.

I was invited to spend the night at the seminary, and I had quite a talk with their leader who was a graduate of high school in Hungary. He was a fine young man and did his best to talk me into joining the seminary. He pointed out that with my "superior education" and my knowledge of Hungary and its customs, I would be a preacher in no time. Perhaps I would even be the leading Hungarian Baptist minister within a few years. I spent a sleepless night considering the offer and the chance.

While I traveled as a peddler, I had listened a few times in different towns to Billy Sunday.[9] I knew that his revival meetings always did something good in every colony, even if the effect was not lasting. I visualized myself as a Billy Sunday in a coal miner edition, going from one Hungarian colony to another, preaching to people, listening to their grief and worries, encouraging them in their struggles, and helping them to solve their many problems. The role was very tempting. And even if I would have had to induce Hungarians to adopt a new faith, a new faith seemed, in my mind, a small price for giving up alcohol.

Then I remembered my mother. I was very grateful for the tact of members of my family for never mentioning to her that I changed my religion. But I realized that if I became a Gentile preacher, I could not conceal my choice from my mother, and the choice would break her Jewish heart. I left the seminary the next morning with a soft spot in my heart for the Baptists who did so much to lessen evils of whiskey amongst Hungarians in America.

I made enough to repay Gyulay by the end of November and accumulated enough stock to risk the expenses of a trip to my old hunting ground, the Virginias. I was greeted by many old customers and friends, but was sorry to note that work was slack. My old customers and friends were moving around a lot, searching for more permanent and steady work, and almost each one asked me if I knew places where such work could be found. I was sorry to note that the miners were completely uninformed. They often moved five hundred miles away on advice from a friend when steady work could have been obtained perhaps twenty miles from their camp. I thought that Hungarian newspapers were very neglectful by publishing births, baptismal ceremonies, weddings, and burials, and yet not informing their readers about, to the readers, the most important aspect of life. . . . work. And I then struck the bright idea that a periodical sheet containing nothing but information about work and working conditions would be welcomed by my customers.

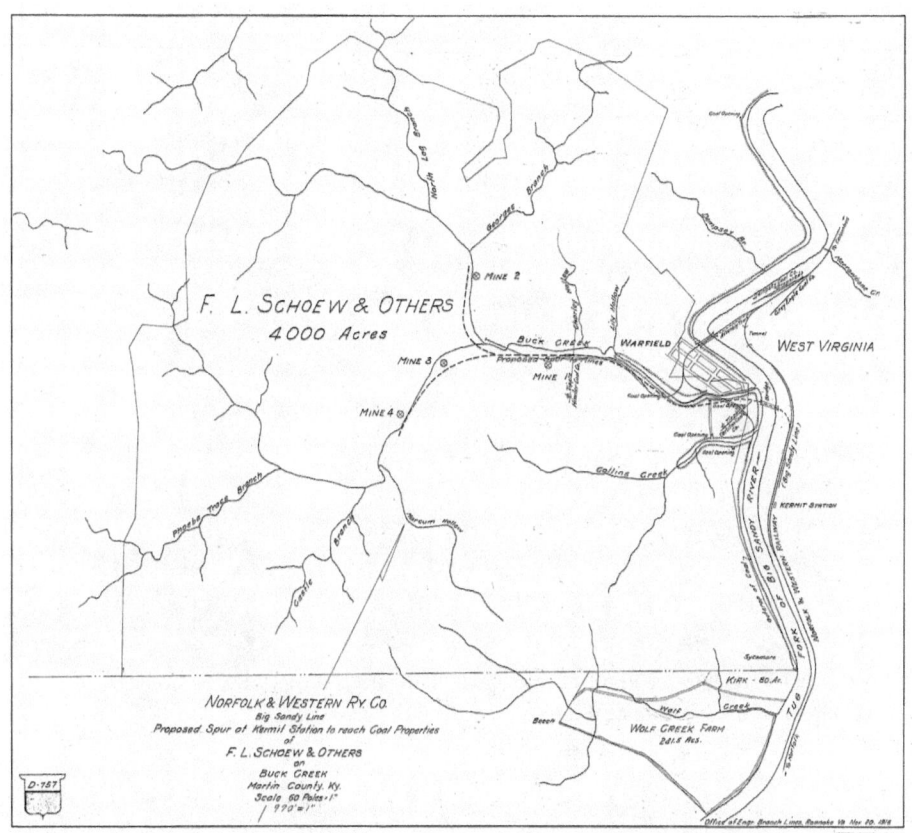

Proposed Spur at Kermit Station to reach coal properties of F. L. Schoew and others on Buck Creek. Courtesy of the Norfolk & Western Historical Society.

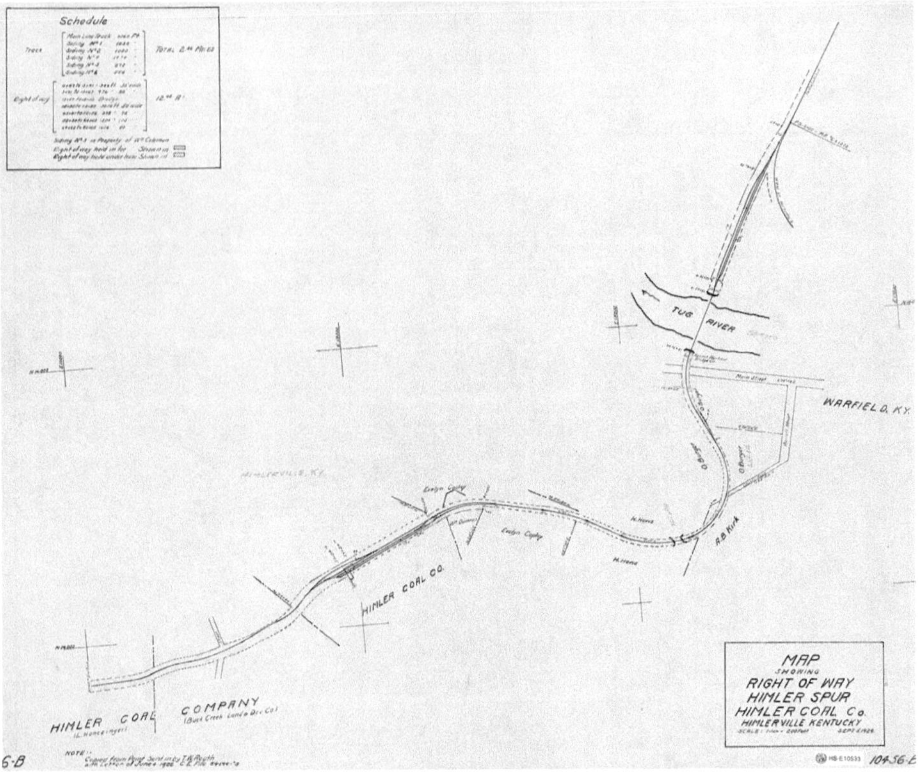

Right of Way, Himler Spur, Himler Coal Company, Himlerville, Kentucky. Coutesy of the Norfolk & Western Historical Society.

S.S. Carpathia. Courtesy of Victor Bill Bodo Estate, submitted by Catherine Bodo Spencer, Mary Bodo Evans, Lisa Bodo Saunders, and Patricia Bodo Sazy.

Construction of Himlerville with the Manager's House in the upper right. George Gunnoe Collection Photos, Marshall University Library Special Collections, Huntington, West Virginia.

Side view Martin Himler's house under construction. George Gunnoe Collection Photos, Marshall University Library Special Collections, Huntington, West Virginia.

Himler Coal Co., Inc., company store under construction. George Gunnoe Collection Photos, Marshall University Library Special Collections, Huntington, West Virginia.

Himlerville residential houses. George Gunnoe Collection Photos, Marshall University Library Special Collections, Huntington, West Virginia.

Himler Coal Co. mine entrance. George Gunnoe Collection Photos, Marshall University Library Special Collections, Huntington, West Virginia.

Inside the Himler Coal Co. coal mine and tipple. George Gunnoe Collection Photos, Marshall University Library Special Collections, Huntington, West Virginia.

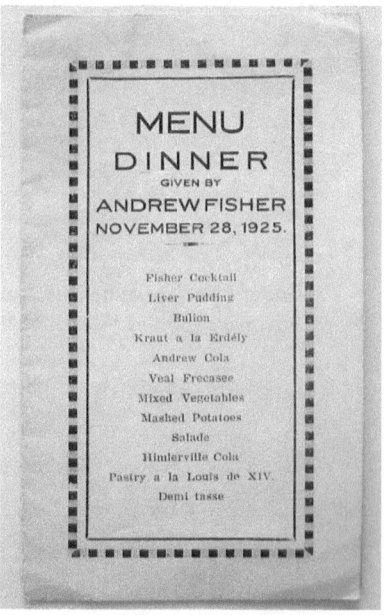

Himler House dinner menu cards. Courtesy of Martin County Historical and Genealogical Society, Inc.

(*Left*) Colonel Martin Himler in his OSS uniform. Courtesy of Himler family.

(*Below*) Himler at his newspaper publishing desk, April 1940, Detroit, Michigan. Courtesy of Martin County Historical and Genealogical Society, Inc.

Camp Marcus W. Orr, Salzburg, Austria. Photo and description from the National Archives and courtesy of the collection of Duncan Bare, Hungarian-American Intelligence Studies, University of Graz, Austria.

Martin Himler and Sgt. Bela Herezeg (Secret Intelligence Central Europe), seated, speak with László Szillágyi (*far left*) and Gabor Vajna, former minister of the interior under Ferenc Szálasi's fascist government. Camp Marcus W. Orr, Salzburg, Austria. Photo and description from the National Archives and courtesy of the collection of Duncan Bare, Hungarian-American Intelligence Studies, University of Graz, Austria.

Ferenc Szálasi and members of his fascist Arrow Cross Party government being held for interrogation at Camp Orr. *Left to right*: Vilmos Hellebronth, Ferenc Omelka, Ferenc Kiss, Peter Hain, Szálasi, Ekno Gombos, Geza Perenyi, Ede Voigt, Lajos Simsai, László Endre (behind Simsai), and Ferenc Kassai. Photo and description from the National Archives and courtesy of the collection of Duncan Bare, Hungarian-American Intelligence Studies, University of Graz, Austria.

```
                    HEADQUARTERS
              UNITED STATES FORCES IN AUSTRIA
                    O. S. S. AUSTRIA
                   APO 777, U. S. ARMY
```

9 October 1945

Subject   : Photos of Historical Interest
To   : 971
              Chief, Central European-Scandinavian
                Section, OSS Washington

    1. Attached to this cover sheet are:

        a) Pictures of various Hungarian war criminals apprehended by SI Hungarian Section and returned to Hungary.

        b) Pictures of Walter C. Langsam in Paris.

retained by WCL

Al
ALFRED C. ULMER, JR ,
Lieutenant, USNR,
Chief, SI

ACU/ay

Cover sheet to declassified photos/Martin Himler interrogations of Hungarian Nazi war criminals. Photo and description from the National Archives and courtesy of the collection of Duncan Bare, Hungarian-American Intelligence Studies, University of Graz, Austria.

MY BOYS:

"My Boys" of the Buck Creek, do you remember "the Old Man"?

The old man who once shared your worries, and your joys.

The old man who was sick when one of you happened to be hurt in the mine, our mine — that was.

The old man who derived so much pleasure in those never-to-be-forgotten days when you were young, watching you enjoying yourselves Saturday nights in the company hall.

The old man who derived so much pleasure watching "Pork Chops" make a home run against "Kenova" or "Portsmouth."

The old man whose heart went out to you when upon the collapse of our grand dream you had to scatter and seek new homes.

And you, youngsters, who are young men now, do you remember the old man dishing out occasional pennies on the store porch?

This is the call of that same old man.

This is his plea for a meeting, perhaps a last meeting, August 30, in Beckley.

I would like to see all of "my boys" once more in my life. I would like to rejoice in your successes, in your families, your wives, children, and grandchildren.

I was denied a family of my own, in my heart my family still only consist of "my boys," the now old boys of Himlerville.

And I would like to see my family "my boys," once more.

Wouldn't you like to see your old comrades, and your old playmates?

Come one, let's all meet in Beckley.

Martin Himler

Letter written from Martin Himler to former Himlerville residents. It was posted in a Williamson, West Virginia, newspaper in an unknown year. Courtesy of John and Helen Younger, former Himlerville residents.

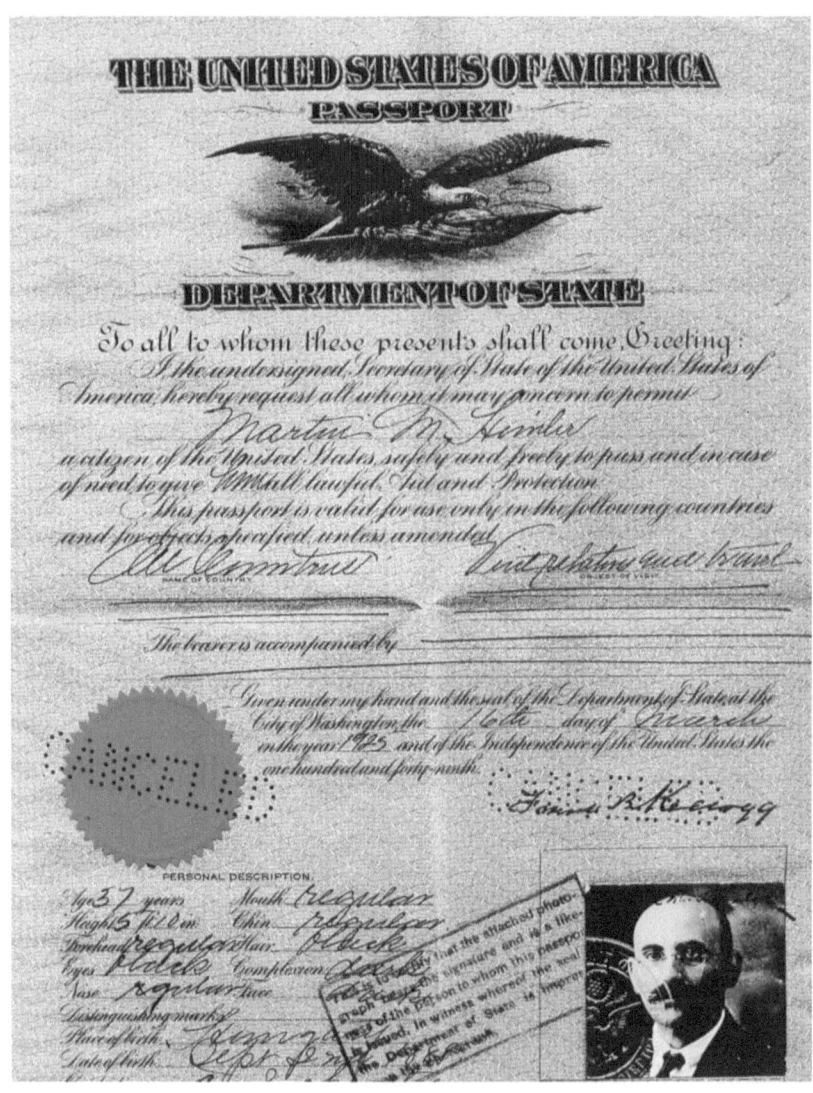

Himler's U.S. passport. Courtesy of the Himler family.

Himler later in life relaxing at his Los Angeles home.
Courtesy of the Himler family.

Himler's ashes are interred in Hollywood Forever's Cathedral Masoleum, Corridor C, Alcove of Devotion, Tier 13, Niche 9, Hollywood Cemetery, Los Angeles, CA. Photos courtesy of Theodore Hovey.

CHAPTER 13

I Become a Publisher

★

WHILE MEN WERE working in the mines during the day, I was just killing time as I waited for my peddling customers to finish their mining shifts. I therefore sat down — in Holden, West Virginia, — to write the first issue of a newspaper to inform Hungarians about work and working conditions in America. I mentioned fifteen coal camps in the first issue of the newspaper.

But somehow, I must have been born carrying the worries and pains of half of mankind for before I finished the newspaper, I wrote editorials enumerating the problems of coal miners, especially the immigrant coal miners. I also wrote a descriptive article about life in Hungary as I saw it on my recent trip there.

When I thought that I had enough material for an eight-page nine by twelve inch (commercial letterhead size) newspaper, I mailed my newspaper to a Hungarian printer in New York who knew me from the *Trailblazer* days. I requested that the printer print one thousand copies of the newspaper and express same — with his bill — to Coeburn, Virginia. I chose Coeburn because my best clients and friends were living at Toms Creek, Virginia, and I wanted to try selling subscriptions at the most favorable point.

The shipment of the first issue of the *Magyar Bányászlap* (*Hungarian Miners' Journal*) arrived in due time. The bill was $17.00. Unfortunately, the printer was raised in Hungary, and the Hungarians write the number 17 as we do the number 77. I then knew so little about printing and printing cost that I did not know whether the bill was for $17.00 or $77.00. I did not know if I could keep the newspaper alive if the bill were to be for $17.00, but I was certain that if I had to pay $77.00 for each issue, the printing of the first issue also would be the last.

The masthead of my newspaper did not state the name of the publisher or editor, but there was an announcement in the newspaper that Martin Himler was authorized to accept subscriptions. I took my newspapers to Toms Creek with high hopes, and I failed to secure one subscription.

Two weeks before my visit to Toms Creek, a so-called picture taker had visited Toms Creek. The picture taker had collected pictures and advance payments for the enlargements of the pictures, but he had failed to deliver the goods. The activities of many seemingly very able salesmen who turned crooks were always beyond my understanding. Salesmen often went around amongst the immigrants and sold the immigrants suits of clothes. The salesmen accepted deposits of $5.00-$10.00 as a down payment, but the clothes, as the pictures, were never delivered. The salesmen would have made more by delivering the article for the original deposit was, in most cases, much smaller than the legitimate profit would have been. Also, an honest salesman would have been able to return to the same place over and over again, just as I did, instead of always seeking new territories and thus sometimes running into a victim previously cheated in a previously visited territory. When a cheating salesman ran into one of his former victims, the victim often took a poke at the salesman to satisfy the victim's anger. There were many such sales crooks operating amongst immigrants and taking perverted joy in bragging about their successes at out-right crookedness.

Anyhow, after my complete failure at Toms Creek, I decided that I was up against insurmountable odds and that I would have to give up the hopeless struggle before I spent too much money and time on the newspaper venture. However, the next morning, I called on one of my old friends, Mr. Chaszar, a small merchant at Coeburn. I told him about my newspaper scheme and my failure. He was enthusiastic about the newspaper, considered it a brilliant idea, and paid $1.00 as my first subscriber. He virtually begged me to not give up on the newspaper. Thus encouraged, I proceeded to Dante, Virginia, a large Hungarian colony at the time. I could hardly believe my own eyes when I counted the thirteenth $1.00 bill collected from the many subscriptions that I sold in just one evening. I knew then that if the printing bill called for $17.00, the newspaper would go on for my business of peddling would enable me to finance the newspaper without any income from subscribers. As a result of my experience at Dante, I knew that I also would have some direct income from the newspaper.

I made some plans on the long return trip to New York, and I prepared a questionnaire to be mailed out to the various local coal mining camps. The questionnaire requested information about the working conditions in the mines, the pay, the permanency of work, and the treatment of the workers. I learned in New York that the printing of the questionnaires would cost me $17.00, and I immediately mailed out fifty questionnaires to that many coal camps. I also prepared material for the second issue of the newspaper in which I would be describing conditions that I had seen in the mines between Coeburn and Dante.

My plans and hopes did not include a career for myself as a publisher, much less as an editor, and I did not expect to make a living by publishing *Magyar Bányászlap*. I merely hoped to serve the miners and thus to increase their goodwill and their patronage to my peddling business. Eventually, I hoped to quit the road and build up a mail-order business.

The masthead of the second issue of my newspaper boldly stated that I was the editor and publisher and that the newspaper was "written for miners, by miners." I felt, as I do now, that this was true. The second issue also carried a one-page ad offering goods of all sorts by mail. But instead of using my name as the editor, publisher, and merchant, I used the merchant trade name of Mátraverebély & Co. which I registered for $.50 at the proper office in New York.

I mailed out three hundred copies of the second issue of my newspapers to local miners in about forty mining camps. The newspapers were carried on the elevated train to the main post office. Since second-class privilege already was applied for, I paid two whole cents for postage. Second-class mail in 1913 was carried for a cent a pound anywhere in America.

As soon as the newspapers were mailed, I rushed around to nearby Hungarian colonies to sell enough goods to make a living and to pay the printer. I worked as hard as I knew how to make both ends meet. Meantime, I prepared the next issue of my newspaper by using a few of the answered questionnaires which had started to come in. I learned that by using the same composition of another Hungarian newspaper, I could publish a standard size, seven-column newspaper at the cost of $22.00, and the third issue appeared in this new form. Since this publication of *Magyar Bányászlap* fell on the third of January, 1914, the issue proudly carried on its front: Vol. 2, No. 1. My few subscribers who gambled a dollar by subscribing to an unknown publication were elated by receiving more than they had bargained for. They wrote letters of appreciation, and a few of the sample copies also brought in some new subscribers.

My one-page mail order ad had far better results than I expected, and this ad was a revolutionary attempt to get business. I stated in the ad that no money would be accepted with any order and that purchases would have to be paid for upon receipt of the goods at the post office. No other mail order house catering to immigrant trade had thought of such an unheard-of method of payment. Further, I stated in the ad that anything not mentioned in the advertisement could be ordered and that the ordered items also would be sent at the best possible prices.

I had to work my head off by writing the newspaper, mailing it, going around in New York to buy goods that I never had heard of before, packing the goods, and carrying sacks full of goods to the post office. My mail order business grew by leaps and bounds, and it was the undoing of the postal department. Fourteen days were required to have the money orders for the COD shipments come in, and although I made nice profits and my outstanding accounts grew fast, I was shorter and shorter in ready cash to buy and ship the new orders.

Since my fingers had been burnt before by running into debt, I would not think of borrowing money. I was too shy to invite any of my friends to join me in partnership. I had to kill the mail order business because it was too good, a silly thing for which I have blamed myself ever since. I omitted the one-page mail order ad from the sixtieth issue of *Magyar Bányászlap*.

Anyhow, I decided that the newspaper would be self-supporting and that eventually, perhaps within a year or so, the newspaper would even support me. As a part of my new plan, I sent out Gyulay to West Virginia to solicit subscriptions. He was to peddle goods on his own, pay his own expenses, and keep 90 percent of the $1.00 annual subscription price. I thought that it would be better to receive at least $.10 than to continue mailing out free sample copies of my newspaper. I was counting, of course, on direct subscriptions for the second year. I knew that with a few direct subscriptions to help me, I could finance the newspaper for a year by peddling around New York and returning home on weekends to attend to the newspaper and to buy goods.

Gyulay's trip was a huge success, far beyond my wildest hopes. There were very few weeks that he failed to mail in about one hundred new subscriptions and within eight weeks, I had over one thousand subscribers and a good coverage of Virginia and West Virginia coal mines. Since I had the income from more than one thousand subscriptions, the growth of my mailing list increased only my work, not my expenses.

There is a Hungarian saying that the appetite comes with the eating. I hired another sales agent and purchased four hundred $1.00 watches at $.40 each. I sent the agent to the much harder coal fields of Pennsylvania with instructions to give away a watch with each newspaper subscription. He was to keep $.50 of each dollar. And I was still to receive $.10 net, just as I did from Gyulay. Of course, this new plan was to increase my newspaper's circulation. Two-thirds of the Pennsylvania subscribers likely did not hope to ever receive a single copy of the newspaper for the Pennsylvania camps were invaded by droves of crooks. However, the subscribers had nothing to lose for they received a $1.00 watch immediately for their dollar.

When the *Magyar Bányászlap* was about four months old, it had two thousand subscribers, more than any other Hungarian weekly in America. But I had to work harder than ever to finance it, write it, typewrite it, and paste the addresses on two thousand copies every week. Plus, I had to attend to my growing correspondence. The newspaper was good to the taste of Hungarian coal miners. I knew their lives and their problems from my coal mining experience and from traveling amongst them. But the best-appreciated columns were the columns containing information about working conditions in the various sections of the country.

I wrote editorials, descriptive articles, short stories, novels, published parts of an English-Hungarian dictionary, and also included information about citizenship in America in my newspaper. I also attacked the many abuses of the mining camps, such as the lack of care for the welfare of the miners by the operators. Some mines had bad ventilation, and some camps lacked good drinking water and decent dwelling houses. There were always abuses by the grafting mine bosses. Such articles were published in both Hungarian and English on the newspaper's front page, and my subscribers did not fail to show the newspapers to their American buddies and the mine foremen.

The United Coal Miners' Union[1] did not reach farther than Illinois, Indiana, and Ohio, but did have some organizations in the open-shop mines of Pennsylvania. *Magyar Bányászlap* became the spokesman of the downtrodden miners of the Southern coal fields. As soon as miners realized that their grievances could be aired without disclosing the source, complaints came in from everywhere and thus gave me material for at least three or four attacks on mine operators for each issue of my newspaper.

I paid attention to Hungary as well. Most of the miners planned to return to the Old Country, and it was safe to attack the system as well as the ruling class of Hungary for the arms of the gendarmes did not reach to America. And the Lord knows that there was plenty to criticize about conditions of the masses in Hungary. Some of my attacks resulted in improving conditions at some mines and increased the prestige of my newspaper to a height never attained by other Hungarian publications. However, at some other mines, my agents were kicked out, or I received letters threatening me with libel suits.

Magyar Bányászlap was self-supporting by the fifth month of publication. I received enough direct subscriptions to cover the printing cost, and I had agents working on straight 50 percent commission without premiums or any other inducements except the merits of the newspaper.

I went down to West Virginia to see the reaction of the subscribers and to personally listen to their criticism. I was elated to note that my attacks on operators were not resented by the more intelligent superintendents. The General Superintendent of the Island Creek mine at Holden told me that any just criticism was appreciated and requested me to visit each of his Hungarian miners to listen to their grievances. The superintendent promised me that every reasonable request would be attended to.

There were close to two hundred Hungarians on the Holden Creek Camp, and almost each had something to say, something to complain about. The complaints were mostly about little things as far as the company was concerned, but little irritations are enough to make men miserable. I thus advised the superintendent to hire a Hungarian clerk, and I was requested to select one for him. I also advised the superintendent to help the men establish a Protestant congregation, and I had to send him a Hungarian Protestant minister.

A few days after my return to New York, I received a check from the Island Creek Superintendent for $191.00 to cover *Magyar Bányászlap* subscriptions for each of his Hungarians. The subscriptions were, of course, for the very newspaper which had bitterly attacked the Holden operation only a few weeks earlier. I mailed the check back with a letter stating that most of his Hungarians were subscribers, anyhow, and that I did not wish to accept favors from coal operators.

The check was returned to me with a nice letter stating that I should consider the check as payment for my expenses for the four days that I spent interviewing the Holden miners. The letter also stated that I was not expected to show any favor to the Island Creek Company in the future, except to criticize them again for any cause that I considered just. I accepted the check.

Another coal mine superintendent placed an ad requesting miners in my newspaper, and I charged that superintendent a stiff price. This was the first paid ad carried by the *Magyar Bányászlap*.

While I was on the trip to West Virginia, my miners called my attention to one aspect of their lives that was neglected by my newspaper. The crooked traveling men, the picture takers, the Hungarian card sharks, and the agents selling lots at the bottom of the sea or stocks in oil companies took advantage of my miners. Cleaning up the dishonest business in the mining regions was right up my alley. I jumped in with both feet.

The minute a crook showed up in any mining camp, I was advised of the crook's presence. The crook's description, as well as the description of his operations, were then published in my newspaper. As popular as I became amongst the miners, I became more unpopular amongst the hundreds of crooks fleecing them. And I was far from being popular amongst the Hungarian publishers and the other leaders. I was an upstart in their eyes, an upstart who talked loud and used a big stick. The mystery of it all was that I did not approach the nearest source of income of Hungarian newspapers. I did not solicit a single ad from Hungarian bankers, steamship agents, manufacturers of laxatives and other patent medicines, or crooked mail order doctors (so-called Medical Institutions) who operated amongst immigrants and the Hungarian Societies.

West Virginia went dry in the early summer of 1914, but taking less than five gallons of whiskey or ordering not over four and three quarters gallons by express was permissible. A few weeks later, my newspaper was full of ads from Ohio and Kentucky whiskey manufacturers. While I was opposed to excessive drinking, I saw no reason to refuse the ads. I needed income badly. When my newspaper was a half-year old, a box on the front page showed that the *Magyar Bányászlap* carried more inches of paid ads than any Sunday issue of any of the large Hungarian newspapers. This fact did not tend to increase the friendship of my jealous colleagues.

Soon after West Virginia went dry, my subscribers called my attention to still another grave abuse to which they were subjected, especially with the prohibition in West Virginia. Mine guards and other deputy sheriffs and justices of the peace considered the immigrants free prey. The slightest excuses had immigrants arrested and scared into handing over all of their money. If the mine guards and other law officers learned that a miner had a few hundred dollars, that miner was sure to be a victim. I started to secure facts and soon was convinced that hundreds of immigrants were being victimized without the slightest cause. The worst spot for victimization was Welch, West Virginia, the main town in the Pocahontas Field.[2]

Coal operators did not try to protect their miners from robbery, and some operators were unaware of the robbery. All coal operators used mine guards because operators feared unionization. Mine guards stripped the innocent immigrant miners of their funds, and the abused miners had to work all the harder to make same money again.

The most notorious man and leader of the mine guard gang around Welch was Squire Hock, Division Chief of the operation of the Baldwin-Felts Agency. Squire Hock had several killings and innumerable beatings of innocent people to his credit. I started my series of attacks on this most-feared Squire Hock by publishing facts and naming victims and amounts. Welch and its vicinity became a beehive.

Squire Hock was considered untouchable due to the support of his agency and the protection of the almighty coal operators of the mines. The town of Welch was so terrorized by Squire Hock that although I received letters of commendations on Squire Hock from ministers and judges, all letter writers requested that I keep their communications confidential.

Squire Hock wrote to me and said that he would deal with me on my next trip to West Virginia, and I invited him to sue me for libel in New York. My miners wrote to me and begged that I stay away from West Virginia, for Hock had bragged openly about his intention to kill me. However, from then on, mine guards in Welch were careful to select their victims from other than Hungarian miners.

A month later, Count Michael Karolyi, then the leader of the liberal party of Hungary, paid a visit to the Hungarian Americans to collect money for political purposes. I supported Count Karolyi, as all other Hungarian newspapers did, and it was a matter of prestige that I go around with him. And to the chagrin of my colleagues, I invited Count Karolyi to the coal fields of the Virginias, partly because it was expected by my miners and partly to have him see that I and my newspaper were his backers in Hungarian America.

I had no choice but to accompany Count Karolyi to the Pocahontas Field, and I do not wish to deny that I was scared plenty. So were my miners. About ten of them attached themselves to me and traveled with me for four days until I left the main line of the Norfolk & Western Railway and the Pocahontas Field. I was especially apprehensive when I had to change trains in Welch for Gary and had to stay in Welch for about an hour and a half, but nothing happened. I left West Virginia a hero in the eyes of the miners. Later, I learned that I was indebted for my life to Mrs. Hock who kept the roving squire locked in their bathroom during my stay at Welch.

While on the same trip to the Pocahontas Field, I called on managers of local national banks. I solicited advertisements and induced quite a few of the banks to engage the services of a foreign-born linguist clerk. When I returned to New York, I sent such clerks to five local banks, and *Magyar Bányászlap* carried the ads of eleven local national banks.

An unprecedented thing happened with the issue of my newspaper that carried the ads of the local banks. The first Hungarian banker to subscribe to my newspaper mailed in his subscription order without being solicited. Since the miners swore by my newspaper, it had a pulling power much stronger than any other Hungarian daily or weekly newspaper. Since *Magyar Bányászlap* was an ever-crusading organ, it was feared and respected by the Hungarian leaders but not liked very much, least of all by the bankers.

I mentioned previously the so-called "running bankers" who skipped with the savings of the immigrants, but there were other lesser abuses, such as over-charges for affidavits or for other necessary services. Once a saloon keeper in Pocahontas, Virginia, told me that while he appreciated the clientele of the Hungarians (his own countrymen) because they always were willing to pay a decent fee for safekeeping of their savings, he preferred the trade of Slovaks and Russians because most of them were illiterates and failed to remember the amounts they entrusted to him from time to time for safekeeping. No, such people as the saloon keeper did not appreciate the ads of national banks or their hiring of linguist Hungarians, on my advice, to cater to the trade of the immigrants.

My peddling days were over. *Magyar Bányászlap* was prosperous. I had an editorial staff in the person of Miss Fulop when, in the summer of 1914, Varhelyi and his partner went bankrupt. Varhelyi did not have a penny nor any inclination to start a shoe-repairing shop. He was totally crushed by this humiliation in the Hungarian colony where he was considered one of the lesser leaders and presided over a group of four hundred.

I had to rent a four-room flat, still in the tenement district on Eighth Street. The front of the tenement served as my office with two black holes for bedrooms for the Varhelyis and myself. The fourth room was the kitchen and dining room, as well as the Varhelyis' living room. I paid the rent for the tenement and $5.00 for my board, and I induced several of my friends to take their evening meals there. This took care of the most pressing needs of the Varhelyi family. Varhelyi attempted peddling paprika in New York but never made over a dollar or two per week.

When my newspaper was nine months old, when it was a very promising business, and when I was riding on the crest of the waves of popularity, World War 1 broke out. My sentiment, at first, was with the Central Powers. My native country was involved in the war. Five of my brothers and as many brothers-in-law, were fighting in the Austro-Hungarian Army, and they were fighting the same Russian army that was used to crush the Hungarian Revolution of Kossuth in 1848.[3] I was deeply offended by President Wilson's coining the infamous name of "Hyphenated Americans" and branding immigrant citizens with the label of implied disloyalty.

A few weeks after the outbreak of the war, my brother, Géza, fell on the Russian front, and this did not increase my love for the Russians. But by the time we Americans entered the war, in fact, with the sinking of the *Lusitania*,[4] my eyes were opened, and so were the eyes of most of the Hungarians in America.

Incidentally, I changed my mind about President Wilson, and I believe that he was one of the greatest men of his age. He conceived the League of Nations[5] before the world was ready to follow his leadership, even though he misjudged the immigrant citizens woefully.

I shared in the war prosperity. Coal companies were filling my newspaper with ads crying for more workers and paying any ad price that I charged. I was not bashful. I visited the mining regions often and on one of my trips, I made a disturbing discovery. I saw several unopened issues of *Magyar Bányászlap* in one of the boarding houses. I was surprised for I was under the impression that readers were awaiting each issue eagerly and could hardly wait to receive the issues. The Mrs. of the house explained that there were twenty-three boarders and that many *Magyar Bányászlap* subscribers in her home. She reassured me that as soon as the miners were home on Thursdays, the first thing that they did was read the newspaper, often before washing up. More interesting articles were read aloud by the first man home to the entertainment of the others. Thus, several of them did not need to open their newspapers. I was still dissatisfied and felt that it couldn't do any good to have unopened newspapers lying around in a house. It would be far better if a reader would be compelled to borrow his neighbor's newspaper when the neighbor was finished with the newspaper.

I decided to do something that very few publishers did or do. I decided to reduce the circulation of my newspaper. As a result of my decision, I did not send notice to renew when I saw that a dozen or more newspapers went to the same post office box. It took me three years to reduce the number of subscribers from approximately six thousand to about four thousand, and I still reached almost all of the Hungarian miners. The income did not matter for my revenues from ads increased faster than I ever had hoped for.

I improved the contents of *Magyar Bányászlap* every week. The best Hungarian writers contributed short stories, written always about the coal miners, and the synopsis for every story was furnished by me. I also started to publish original funnies.

I was the most independent publisher in Hungarian America, and I used my independence for good purposes. I did not care a tinker's damn whether Hungarian businessmen or societies were advertising in my newspaper or not, and I criticized businessmen, priests, ministers, and leaders of the various societies freely. They disliked me immensely but were unable to touch me for the miners were out of their reach, and my advertisers were coal companies, breweries, national banks, and local merchants on the coal fields. Although, twenty years later, the Hungarian businesses and societies had a field day!

As a result of a little prosperity and prestige, I had several small thrills in those days. Once, in the company of Count Karolyi, I actually hired a taxi for my first ride in such vehicle. Up to then, a taxi was completely out of my class. I started to travel in Pullman cars, and once—remembering the skin of a grapefruit I picked up when I was a hobo and was very hungry—I went into the dining car for breakfast. Dining cars were a luxury not practiced by Hungarians before World War II.

I began to drop in for a cup of coffee in the afternoons at the Café Monopol or Café Royal on Second Avenue. I met other Hungarian intellectuals there and also met the newspapermen working on the Hungarian daily of New York.

When I was still doing some peddling during the early days of my newspaper, two coal miner friends gave me a $600.00 investment in my newspaper business. My friends did not expect me to pay any profits to them; the $600.00 was offered only to keep my newspaper going. However, in 1915, I paid the first dividend of 12 percent to my stock holders, and thereafter, 12 percent annually. Twenty years later, my friends agreed to sell their stocks at a 100 percent profit.

I was put up for the night by one of my former investors when I made one of my peddling trips in 1915. I bunked that night with my investor's son, a young man about sixteen years old. I noted that there were healed wounds on the young man's body, and I learned that the roof had fallen on the young man when he was a kid. I persuaded his parents to let me take him with me to New York and send him to school at my expense.

As my income increased, I selected two more of such boys and sent them to the private school maintained in New York by the Young Men's Christian Association (YMCA). I was hopeful that some of the young men might be the future leaders of the Hungarian coal miners. I was disappointed in this hope for all the young men became businessmen, but I derived much joy in having them with me in New York.

Every now and then, I was invited to be godfather to a child of some of my friends. This was an honor that I could not, nor did I wish, to refuse. Several of my godsons are closely attached to me, even now. Indeed, I worried about them during World War II as if they were my own sons.

Beginning with the first issue of *Magyar Bányászlap*, I called my readers "miner brother" which sounds a lot better in Hungarian (Bányásztestvér). I encouraged the readers to call each other the same way in order to promote fraternity amongst them. No Hungarian miner to this day would think of addressing another Hungarian miner by any name except "miner brother."

I trained my miners to call on my newspaper for any assistance when needed, and they certainly took advantage of the privilege. If a miner were crippled in the mine or if his hog were killed by the railroad, he invariably turned to me for advice and assistance, whether he was a *Magyar Bányászlap* subscriber or not. If miners donated money to their "miner brothers" in need, I immediately sent the donated money to the needy miner. If a miner requested some information from me and was thoughtful enough to enclose a stamp for the sending of my answer, I always returned the unused stamp with my answer. If a miner were in real trouble, such as being innocently charged with a crime, he turned to me for help. If the cost of his defense were beyond the financial ability of my newspaper, I collected the necessary funds from the other miners.

There is one case that I have to mention on these pages. The body of a murdered man was found in Pike County, Kentucky, and parts of his scalp were missing. And, in the year 1929 of Our Lord, it was decided by the prosecuting attorney and by the population of Pike County that the unknown man was victim of a ritual murder by Catholics. The Catholics supposedly needed his scalp for their dark ceremonies. Four foreign-born miners were arrested and charged with the crime. Two of them were sons of my subscribers and friends, and the accused miners were almost lynched before the trial. Local newspapers discussed the case and the various aspects of the ritual murder and thus inflamed the public to a dizzy pitch.

I retained the leading attorney of Prestonsburg, Kentucky, who later became Congressman A. J. May. Mr. May recently had been convicted for using his influence for and accepting bribes from war contractors. On Mr. May's advice, we retained all of the other local attorneys to prevent the prosecuting attorney from doing the same.

Never will I forget the trial of the miners—attended by more people than the courtroom held—and the selection of the jurors. We retired to talk over the available members of the jury. One of the attorneys said that he would not like X juror because he sent up X juror for moonshining during the attorney's term as prosecuting attorney. Another attorney said that we should include Aunt Y because she recently loaned $200.00 to one of her nephews, and so on, until each prospective juror was discussed.

Noting the care with which the case against the miners was prepared, I was afraid that the defendants would be acquitted and that certainly would have meant immediate lynching. On the other hand, I knew that if the miners were convicted, the Court of Appeals would throw out the case. Luckily, the sentence was death by hanging, and the case was disposed by the Court of Appeals, as I expected.

I was a prosperous publisher by 1916. I had no other worry except my worry about members of my family in the service of the Austro-Hungarian Army and the outcome of the war with the fate of mankind in the balance. I was the accepted leader of about sixty thousand coal miners in the American-Hungarian community of about half a million, and I was a man not to be ignored in community affairs.

The money of the Central Powers was selling at a daily more unfavorable rate of exchange about this time, but the Hungarian bankers were still charging $20.00 for 100 Kronen when the same was obtainable for about $15.00-$17.00 in Zurich and New York. One of the Hungarian bankers in Pittsburgh was more honest than the rest and started to offer Hungarian money at the prevailing rate.

Under the leadership of the Transatlantic Trust Company, a large New York bank owned by the Hungarian government, many newspapers published ads ridiculing the "cheap money" of the Pittsburgh bank. Finally, when newspapers became afraid to accept more ads from the ridiculing bankers, even though the same newspapers were still accepting ads offering war bonds and money orders at $20.00 for 100 Kronen, a new daily newspaper was financed for the purpose of destroying the Pittsburgh banker and maintaining the fiction of good vs. cheap money.

I attacked the Transatlantic Trust Company, the leader of the group of newspapers attacking the Pittsburgh banker. The Transatlantic Trust Company was a steady advertiser in my newspaper, and I placed their ridiculing ad into the middle of an article that I wrote against the Transatlantic Trust Company to demonstrate that their patronage did not prevent me from publishing the truth about their practice. The Transatlantic Trust Company convinced other daily newspapers to attack me, but all of the newspapers were afraid to defend the dishonest practice. Thus, several small private bankers fell in line with the Pittsburgh banker.

I attacked the Transatlantic Trust Company again and again until, finally, they had to sue me. They charged me with libel and with conspiracy in cahoots with the few honest bankers. I was invited to the office of the Transatlantic Trust Company's prosecuting attorney, and I was warned to cease attacking the large financial institution or suffer dire consequences. Of course, I published the interview with the attorney and attacked the bank and the gang all the stronger. I was arrested and put under $500.00 bail, but I continued to attack the Transatlantic Trust Company and its unholy practice.

The stupid Austro-Hungarian Consul General of New York (not my godfather) was induced to come to the defense of the Transatlantic Trust Company, and he issued a statement under his official seal. The statement acquitted the Transatlantic Trust Company of all charges although the bank was still cheating its clients. I attacked the Consul whose prestige was the last word and his word the law in Hungarian America. The daily newspapers were still silent and mindful of the fat contract of the Transatlantic Trust Company, but about ten weekly newspapers joined me in my fight. When I was afraid that my attacks would tire the readers, I published cartoons instead of articles to ridicule the untenable position of the Transatlantic Trust Company and the Consul.

Finally, every bank and banker had to come to terms. They had to sell Hungarian money at the ever-lower prevailing rates. My prestige was unexcelled amongst Hungarians, and my services appreciated by my readers. Although I was arrested seven to eight times after the publication of each libel, I never had to stand trial.

When the United States entered World War 1, the Transatlantic Trust Company was liquidated by the custodian of the properties of alien enemies, and its president was interned. The Federal Bureau of Investigation (FBI) knew something that I did not. German and Austrian spies in the United States were paid through the Transatlantic Trust Company.

I certainly had a good time of it until the United States was about to join the Allies. War was declared by the United States, and Hungarians, both those who had become United States citizens and those who still were aliens, were scared to death, myself included. We knew how enemy aliens were treated in European countries, not excluding England, and we were ready to share their fate in a concentration camp, especially after having been branded by President Wilson as "hyphenated Americans." Of course, we took it for granted that German and Hungarian newspapers would be suppressed at once and that if the rank and file were sent to concentration camps, dangerous people, like editors and newspapermen in general, priests, ministers, etc., would be imprisoned for the duration.

In sober moments, Hungarians did not believe that millions of people could be sent to camps, and we were speculating on the lesser evil, the immediate unemployment of half a million Hungarians—most of whom were aliens. This vast army included myself and my newspaper staff of three. The balance of my checking accounts showed about $5,000.00, and I assured my associates that as soon as *Magyar Bányászlap* was suppressed, I would divide the money between the four of us so that each of us could face the dark and unknown future with equal chance. We made plans to move together and live together to make the money last longer. The very best that we dared to hope for was unemployment of Hungarians.

When the declaration of war came, foreign language newspapers received the order to submit translation of every article or advertisement pertaining to the war and to submit such translations to the postmasters when mailing the newspaper. Within days, George Creel's[6] United States Committee on Public Information Office was organized to keep in touch with the foreign language press, and regulations were issued regarding movements of German aliens. A few days later, distinction was made between German and "friendly alien enemies," and we knew that Hungarians thus would not be disturbed in any manner. Few Americans realized, or would ever realize, the effect of this distinction on the Hungarians.

The fact that America considered Hungarians loyal to America sold Hungarians on America for good. A few weeks later, I registered for the draft. Whereas a month earlier I would have done that with mixed feeling, by now I was eager to be drafted, eager to fight for America, the country of freedom and tolerance.

During the days when America's entry into World War I was imminent, I worried about the possibility of being drafted instead of interned. I still had four brothers and five brothers-in-law in the opposing army of the Central Powers, and the possibility of meeting them on the battlefield was a nightmare that I did not like to think about. When I presented myself to register for the draft, I was sent home for my citizen paper. I was bald headed, my face was decorated by a long moustache, and clerks did not believe that I was below the age limit of thirty years old. While waiting for my call to arms, I made arrangements that, in my absence, Miss Fulop was to carry on my newspaper business as I had stated in my will.

Instead of *Magyar Bányászlap* being suppressed during war time, it was more prosperous than ever before. But I was getting restless for the newspaper had become work. There was no worry about finances and business went on at an even keel, much too even to suit me. I also was tired of New York.

CHAPTER 14

I Become a Coal Operator

★

Having been raised in a village, I never felt quite at home in any large city, and I always felt the lack of the fellowship of a small community. Finally, I decided to leave Miss Fulop in charge of *Magyar Bányászlap* and look for some employment on the coal fields until my draft call came.

August, 1917, brought me a visit from two old friends, Joseph Harkel and Louis Hajnal, both coal miners at Powhatan, Virginia.[1] Joseph and Louis had a strange request for me. Both of them were mine sub-contractors, and both had about twelve men working for them. Joseph and Louis had decided that they could do better by buying a small coal mine for themselves, and they wanted me to assist them in finding such a mine. Within three days, I was bewildered by the problems of a coal operation, and I realized that I knew less then nothing about my duties. Luckily, the coal was sold before we took over the mine for the Norfolk & Western Railway, and I did not have to worry about that phase of the management of a coal mine. My associates knew quite a bit about the inside operation of a mine of that size, a mine worthy of producing about fifteen tons of coal daily.

I purchased books, visited neighboring operations, pumped salesmen for all they knew about the various phases of the mining business which, in many cases, was a lot more than I knew, and slowly, I learned the job and the griefs of a coal operator. Of course, my friends and I took the mine with the intention of expanding, and I had to learn something about the capacity of power plants and the limited capacity of our own power plant. I had to learn about the merits of various cutting machines, the electric locomotives vs. the mules in the mine, and the difference between the storage battery locomotives and the straight ones. I had to learn the different kinds of lubricators and the different requirements of each piece of machinery for such lubricators. I had to learn about the construction of an electric motor, the functions of the armature in such motors, and the many-sided activities of the men inside and outside the mine. I also had to learn the buying of hundreds of supplies for the mine.

We had a small commissary, and I had to buy shoes, candies, blasting powder, lard, patent medicine, and the Lord knows what else until I secured the services of a competent Hungarian store manager. I had to keep books, complete the dozens of forms required by the wartime regulations of the government, keep the payroll, and attend to the finances of the company. I had to assure our native workers that the Hungarian owners, themselves coal miners, would not freeze the native workers out of work but would enlarge the operations to take care of as many miners as possible. I had to have the dilapidated dwelling houses repaired, and a few new ones built. And I still had to contribute a few articles to the *Magyar Bányászlap* every week.

I seldom slept over four hours within each twenty-four. If nothing else, the worry that I would be called to military service before I trained my young assistant, a Hungarian boy whom I took away from my newspaper, kept me awake.

The Himler Coal Company's Board of Directors was composed of myself, my assistant, and three coal miners. All of the Hungarian miners were stockholders in our company's co-operative venture. I did my best to have all of them free of the problems of even a small operation before I was called for service. And to take care of any possible free time I might have, our United States government's fuel administrator,[2] Harry A. Garfield, ordered me to visit various mining camps and make speeches to the Hungarian miners about the importance of loading clean coal.

Our mining company received our mail at the small post office of Lenore, West Virginia, about two miles away, and I wrote to the United States Post Office Department about the possibility of establishing a post office at Ajax, West Virginia, the site of our mine. To my great surprise—and I won't deny that it was a pleasant one—I was appointed Postmaster at the post office of—Himler, West Virginia. It was the custom to name post offices after the name of the coal companies. The compliment was not intended to me but to our company.

I BECOME A COAL OPERATOR

While living at Himler, West Virginia, I learned much about the life of the "one third of our nation" mentioned so often by President Roosevelt. The natives of West Virginia were hillbillies, ignorant and illiterate. And small wonder. The mining operation was too small to keep a physician, and the health of the miners and families was attended to by an old doctor at Lenore. We had to keep him on, although I soon learned that originally he was a veterinarian and cured men only by the authority of a certificate received from the West Virginia State Board of Health decades earlier.

My call from the draft board came on a spring day in 1918, and I had to present myself in New York for physical examination. Our country did not then need an army of ten million, our government was too choosy, and because of my flat feet, I was taken only for "limited service" with the statement that I likely never would be called. Limited service meant, at that time, service in one's own trade, and no coal operators or foreign language editors were needed overseas. With the likelihood that I never would be called, I started to plan expansion of our Himler Coal Company mine.

A stockholders' meeting with about one hundred Hungarian coal miners present increased the capital stock of our small company to $100,000.00. The money was secured within a few days and was spent on new buildings and new equipment. Our board of directors also declared and paid the first dividend of 6 percent. But while expanding and investing more in our company, I knew the defects of our mine, and I planned to get rid of same.

Our coal was dirty and contained several laminated veins that, in normal times, had to be washed at extra expense. Since we were living in wartime, our ownership consumers were glad to get any coal and, in fact, some of our miners did not mind to load a good bit of the laminated coal stored up years before in working places. However, loading coal with laminated veins brought complaints from the railroad.

Our equipment was old, our power plant fit to be junked, and I knew that with the close of the war we would have a very hard sled. The mine was a war baby. The Eastern Sales Agency of Columbus, Ohio, specialized in buying up white elephants and unloading them on venturesome operators at a fat profit. Innocent babe in the woods that I was made Himler Coal Company one of their victims.

I secured a lease on two thousand acres of virgin coal land about ten miles from Ajax on the main line of the Norfolk & Western Railway, opposite Kermit, West Virginia, and on the Kentucky side of the Tug River.

I planned to build there a modern mine with new equipment and a modern camp. I was dreaming of Roebling, New Jersey.

The miners of our corporation who were present at the next stockholders' meeting voted enthusiastically to increase the capital to $1,000,000.00 in order to build the new mining town on the Kentucky side of the Tug. All of them wanted to work in their own mine and, of course, they approved the idea of a large and modern mine and mining town.

I looked around for a customer to take the Ajax mine off our hands, but the old reputation of the mine was bad, and experienced coal operators knew its faults. Finally, I induced a Mr. Batta, one of the leaders of the Slovaks, to follow our example and organize a co-operative mine for the Slovaks. To demonstrate to him the merits of the mine, I subscribed $5,000.00 in his contemplated company. We had made about $25,000.00 profit on the sale and could afford to write off the $5,000.00. Coal operators of the field who were skeptical about my ability as an operator and about the nonsense of coal miners' having their own mine told me later that I established myself in their esteem with this sale. As it happened during the dizzy post-war days, coal sold at a much higher price than the wartime ceiling price, and the Batta Company did not lose very much when finally they gave up the struggle.

When the sale of the Ajax mine was completed, I moved over to Warfield, Kentucky, about a mile from the contemplated mine, and started to sink the shaft for the new mine. The Tug River had to be bridged and a two-mile spur had to be built to the Norfolk & Western Railway. Railway engineers made the survey for the bridge which was to be ready in four months at an estimated cost of $125,000.00. Everything else, except the opening of the shaft, had to wait for the completion of the bridge for we had no way to haul heavy equipment across the river.

I had enough money, about $700,000.00 of our capital was subscribed and paid for, and thus plans were made for our new mining town. The town would be a modern town with bathtubs in every miner's home. A small state bank was organized at Warfield with a capital stock of $25,000.00 to assist the company in days to come and to serve the coal miners. The bank was, of course, owned by the coal miners and like the coal company, the bank was called Himler State Bank.

World War I Armistice[3] was then declared, and I was in New York waiting for the news. I was watching the delirious crowd on Times Square with

Miss Fulop when I told her of another of my schemes. Knowing that tens, if not hundreds of thousands of Hungarians would rush home as soon as travel was possible, I was thinking of a steamship agency and a small hotel to take care of the Hungarians while they waited in New York for their turn to return home.

Miss Fulop was all aflame at once. Whatever dreams I had had during the previous years, including my ambitious plan to apply my Roebling experience to the new camp, were always concocted with her taking her share in them. Nothing would satisfy her but to start shopping at once for a suitable hotel building, and we left Times Square for the East Side in search of such a place.

We passed by a vacant building on Fourth Avenue at Tenth Street and called on the agent about the rent and lease. The rent was reasonable but, of course, I had no intention of jumping into the new venture at once. I knew that it would be months before peace treaties would be signed.

The agent was persistent and to escape him, I gave a deposit of $20.00 as a tuition fee for his information while I thought about my plans. When Miss Fulop and I returned to the street, I turned angrily on Miss Fulop for forcing me to lose $20.00. I was shocked to see that she meant business and considered the business closed.

It was of no use to point out to her the fact that it might be long months before business could be started and that I did not have any cash to invest at the moment. She was adamant. In fact, she became furious and called me a coward for trying to lose $20.00 without any attempt to go after my money. She begged me not to leave the city before I at least made an attempt to start the ball rolling.

I got good and mad and borrowed the $2,000.00 rent for a few days from one of my friends in New York. I secured a manager for the building, a man who was for years the best agent of the *Magyar Bányászlap*. I then furnished the house the following day.

The manager took charge, secured tenants for the forty furnished rooms within a few days, and although we had to wait for over a year to open the Miners' Home, we did not lose any money while waiting. The manager broke even.

When I returned home, I accepted $100.00 or at most, $200.00, from my best friends amongst my miners, plus $200.00 from Miss Fulop. The loan from my New York friend was then repaid by the new hotel company.

When travel to Europe became possible, the Miners' Home became a gold mine. We paid a 100 percent dividend within two months, and a few months later, another 200 percent dividend. We had about $30,000.00 in the treasury after one year and could have distributed another 1,000 percent profit on our investment. However, I noticed that the Hungarians had stopped going home because they did not like to be Romanian and Czechoslovakian subjects according to the disposition of the stupid Trianon Treaty,[4] and I proposed that the Miners' Home liquidate.

The manager begged me to wait and said that he felt confident that travel would increase again. Since stockholders already had received their investments three-fold, they supported him. The manager had a fat job. The Miners' Home kept going at a loss until all the $30,000.00 was gone, and the place was closed up.

Two months after the Armistice, I was on my way to Hungary. I did not know anything about my family, except that Géza fell in the early days of the war. I was anxious to see those who were left.

Passports were issued readily, but I was commissioned by the "Hoover Kitchens" organization to survey the situation of Hungary and to make recommendations about opening several kitchens to feed the hungry children there. I received my passport for that purpose.

Travel was very difficult in Europe. Upon my arrival in Zurich, I was told that the next train for Vienna would not leave until four days later. I finally obtained passage on a French diplomatic train going to Warsaw in exchange for a bribe of $200.00 from me plus the price of the ticket. There were further delays in Vienna, and this world-famous "gemütlich"[5] city prepared me for the sights I was to meet in Hungary. Thousands of people were standing in the snow in front of Vienna's Grand and Bristol Hotels, hotels frequented by foreigners and black marketers. The thousands were begging for a piece of bread or a piece of rag. The population certainly was worse off than after the Second World War in 1945.

I tried to measure the extent of the black market and offered $10.00 to the porter of the Grand Hotel for a piece of bread in addition to a small slice of black bread received by each person. Two hours later, my $10.00—a fortune in those days of inflation—was regretfully returned by the porter. He was unable to secure bread.

Finally, on a cold December Sunday morning, I reached Budapest. I was afraid to see my brother. I went to the Ritz Hotel to get the room for which

I had cabled from New York, but I was told that the hotel was housing the various Allied missions and no rooms were to be had. For $10.00, the porter secured me accommodations in the Hotel Royal, and, about 10:00 A.M., I went to look up my brother, the cabinetmaker. I was dreading the impact of the news about my family.

My brother and his family were well-off, but I learned that my oldest sister, with whom I boarded when a student, died of privations. My oldest brother had lost his eyesight during the war. One of my sisters was left with three small children because her husband was "missing," and my youngest sister, Regina, was left a widow after two months of married life. She lived with my mother in Mátraverebély. As a result of breadwinners' being in the service and of the destructions of the war and the Bolshevik Regime,[6] the Hungarian people were impoverished. Some of my remaining sisters were in dire need.

My train was not to leave Budapest for Mátraverebély for three days. Automobiles and other valuables had been cleaned up in Budapest by the Romanian army. I was desperate to proceed to my mother.

One of my distant relatives, the owner of an automotive garage, succeeded in hiding a car during the Romanian occupation, and he was willing to rent the car to me for the trip to Mátraverebély. I left the same day for Mátraverebély in the company of my sister-in-law, her baby, and my oldest nephew, Andrew Fay Fisher. Andrew was the son of my dead sister, and he was employed by an insurance company in Budapest.

The automobile, one of the very few left in Hungary, created a sensation everywhere, and we were stopped at almost every village by the gendarmerie. As I learned, two of the outstanding Socialist leaders were kidnapped in a car, and their bodies had been fished out of the Danube on the previous day, so the authorities were on the look-out for the car. Later I heard, and in 1946 I proved to my own satisfaction, that the Socialist leaders were kidnapped and killed at the instigation of Hungarian Regent, Miklós Horthy.[7] The sham investigation was ordered to quiet down the population.

We could not make Mátraverebély the same day and spent the night in my brother's home in Pásztó. A messenger was sent to tell my mother that I would arrive the next morning, thus saving her from a last-minute shock.

I can't attempt to describe the meeting with my mother and sister in Mátraverebély. Just as I was uninformed about my family, they had not heard anything from or about me for two years. They took it for granted that I was serving in the American army. After exchanging information and learning the conditions of the needy members of the family, I distributed about $2,000.00 between them. This was a very large amount, enough to enable everyone to start life anew. I was hit for $500.00 by the village for a new bell to replace the one that was taken for guns, and a bunch of men called on me to extend their thanks for the assistance given them during the war.

Forty of the boys from Mátraverebély were taken prisoners by the Russians in the early stages of the war, and as soon as the boys contacted me through the Red Cross, I sent them a few dollars. An additional monthly remittance was included in the budget of the *Magyar Bányászlap*, and the money was sent automatically to the boys by Miss Fulop until the end of the war. Meantime, Russian rubles became all but worthless, and the prisoners were the richest boys in the large camp where they were confined due to the regular receipt of my few dollars. The boys were very appreciative of my generosity, and I had good reason to remember the praises heaped on me in — 1945.

When I returned to Budapest, I opened an office to mail out the four thousand letters that I carried from my newspaper subscribers, letters telling the first news about the subscribers to their families. Each letter requested each recipient to mail an answer to Budapest so that I could take the answers back to the States. I was interviewed by the local newspapers since I was the first Hungarian from the States to return to Hungary following World War I. My Budapest office was besieged daily by hundreds of people seeking news about their loved ones. I became a national celebrity, especially when the people learned that one of the objects of my visit was to make preparations for the opening of free kitchens.

One of my friends, a New York Hungarian banker, arrived in Budapest after I did, and we traveled together to survey the post-war situation. We traveled by the same car that I rented the first day, and we had an order written personally by Regent Horthy to request that all Hungarian authorities extend all assistance needed or called for by me. I bribed the mistress of one of the Allied officers to relinquish her room in the Ritz Hotel for $100.00, and I paid 400 Kronen a day for the room, the equivalent of $.16 in those days of inflation. Jewelry, rugs, and objects of art were offered at similar low prices, but I felt that I would be robbing graves if I were to profit by the misery of the people, and I did not avail myself of the opportunities.

One day, a mutual friend of the banker and myself rushed into our rooms and brought the glad tidings that we were invited to tea for the same afternoon by the Prime Minister. The bearer of the wonderful news was a former commercial attaché of the Hungarian government in the United States but was recalled two years before the war.

I told him flatly that I would not attend tea because I was to leave for Mátraverebély the same day, and there would be no other train to Mátraverebély for three days. Also, I did not care to sit at the same table with any member of the government of the "White Terror" Regime,[8] much less accept his hospitality. In no case would I have been willing to disappoint my mother, and even if the Mátraverebély trip had not been contemplated, I would not have accepted the invitation.

The banker was quick to take advantage of my stand for he was just as reluctant to attend for the same reason that I was. He said that he would not go to tea alone. Mr. Zerkowitz, the ex-attache, was desperate. He pointed out that an invitation by the Prime Minister was a command and that in Europe, it is impossible to refuse such an invitation. Further, Mr. Zerkowitz would be disgraced by our decline of the invitation and would become a persona non grata if we failed to appear after he engineered the invitation. I told Mr. Zerkowitz that he did not consult my friend and me as to the timing of the tea and, while I was appreciative of his attempt to please us, under no circumstances would I ever sink low enough to pay my respects to one of the bloody-handed terrorists. Later I learned that Mr. Zerkowitz did not suffer and did not lose face. It was simply decided that we were uncultured American peasants below the understanding of such honors.

A few months later, the Prime Minister was kicked out of office and decided to visit the United States to collect money for charitable purposes. Of course, his trip was financed by the state. He did spend a few weeks here and collected some money, but he forgot to pay his bill when he checked out of his New York hotel. The bill included some pairs of pants delivered COD and paid for by the management of the hotel. When American Hungarians paid his bill to save the good name of Hungarians, I had the pleasure of chipping in $14.00, the price of his pants.

Two months later, I returned to the States with inevitable poppy seed biscuits in my handbag and with the last whisper of my mother that our meeting was to be our last. This, however, was not the case.

I found the activities in Kentucky in bad shape upon my return. When the pillars of the railroad bridge were sunk, quicksand had been revealed at the depth of fourteen feet, instead of the expected rock bottom as was indicated by the surveyors of the railroad company. I was greeted with the news that Himler Coal Company would have to drill down forty-two feet to reach solid bottom. This meant that bridge costs would increase from $125,000.00 to $450,000.00, and the time necessary for bridge construction from four months to a year and a half.

I was in a very precarious situation. My successes were not appreciated at all by other Hungarian leaders. Jealous publishers, offended priests and ministers, and especially the multitudes of petty crooks I drove out of the mining regions, started a loud whispering campaign amongst my scattered stockholders, now numbering over one thousand miners, that the railroad bridge delay was only an excuse. The campaigners said that I was stealing hundreds of thousands of dollars, that the mine never would be built, and that I would skip with most of the money to Europe.

Parts of these stories were believed even by well-meaning Hungarians for a million-dollar undertaking was such a daring, unheard-of venture that no one else even would have dreamt of attempting it. By that time, *Magyar Bányászlap* was fighting the radical Hungarian newspapers tooth and nail for the erstwhile Socialist daily newspapers had become Communist organs. And the radical newspapers attacked my concerns within a safe margin of libel but with enough venom to create uneasiness amongst the miners.

And with this atmosphere of suspicion, I knew that I would need another $500,000.00 to complete the mine and the mining camp. The delay on the bridge meant a longer operation at the mine shaft where development already was started and with the coal stored around the shaft.

To make my task harder, the buyers' strike and the recession of 1920 set in, and miners were not working over two or three days a week, not conducive to inducement to buy more stock. And to boot it all, our operation was opposed by an antagonistic local population.

CHAPTER 15

Himlerville, Kentucky

★

To increase the jealousy and the enmity of the Hungarians of America, a wide-awake physician of Warfield, Kentucky, applied for a post office for my prospective mining camp without my knowledge. The physician's application was made in order to have his wife appointed postmistress. One day, I was surprised to hear that the Post Office at Himlerville, Kentucky, was established.

I was pilloried in the radical Hungarian press as "the vain adventurer, Martin Himler, Himlerville, Martin County, Kentucky." The press took it for granted that I engineered even the naming of the county after myself. As it happened, naming the post office the Himlerville Post Office all but broke my heart for I planned to name the town — Mátraverebély.

Childish? Perhaps.

But I wanted to more than vindicate my mother's faith in me, and I visualized her pride and happiness if, and when, she received letters in Mátraverebély, Hungary, from Mátraverebély, America.

The population of Martin County did not appreciate my invasion of their backward bailiwick. Martin County was one of the poorest counties in Kentucky, and to this day, the only railroad in the county is the one that I built for them. The county seat, Inez, is located in the center of the county about ten miles from the nearest railroad station in Kermit, West Virginia. Mail was carried along this road from Inez to Kermit by hack[1] in the summer time and on mule's backs in the winter time when the road was impassable. Our mine was located two miles from Kermit on this road, and our mining town three miles from Kermit.[2]

The backward, mostly illiterate hillbillies were farming very small patches of land and raising almost exclusively corn for both bread and moonshine. A few hillbillies worked for a gas company that had five wells in the county, and a few worked occasionally for a small timber company.

There were three homes equipped with bathrooms in Inez and two in Warfield, a town located between Kermit and Himlerville, and this was the realistic measure of the standard of the Martin County people. Many of the log cabins scattered on the hillsides had but one opening that served as both door and window.

School was maintained for only four months during the summer season when students could walk to school without sinking into knee-deep mud holes. Three years after the establishment of Himlerville, the Kentucky public school superintendents were ordered by the state board of education to attend and graduate from high school, and the standard of the teachers was in line with that of the superintendents. The major part of the in-bred population was of very limited intelligence, but about one-fourth of them were blessed with a native shrewdness, a testimony for the original fine stock of the population.

In the eyes of the local community, Himlerville represented two evils to be fought tooth and nail. The Himler Coal Company was "Big Business" by the standards of Martin County and thus represented all that was undesirable in Wall Street. On top of that, we were "furriners," the first from a distant land to invade the sacred soil of Martin County.

Resigning myself to the unavoidable delay in the building of the bridge, I had the necessary heavy machinery dragged across Tug River, and I started to build both the railroad on the Kentucky side and the heavy steel construction of the tipple. The tipple is the building containing machinery to lift the coal from the shaft and separate it into three or four grades according to sizes that are then loaded into the railroad cars.

Himler Coal Company had to cross several small parcels of land with the railroad and before I knew it, I was faced with about fifteen injunctions by the land owners. The owners did not know that their fathers or grandfathers also sold rights-of-way on the land's surface when the coal rights were sold long before the current owners were born. In spite of Himler Coal Company's indisputable rights of way, I offered the current land owners twice the price of the strips of land that I had to take. But to the landowners, the issue was often a matter of principle and a fight against Wall Street and the "furriners."

The attorney for the current landowners was well aware of the hopelessness of the legal aspect of the cases, but the cases themselves were not hopeless by any means. Once I listened to a campaign speech by Circuit Court Judge Bandy, and I was astonished by his bragging that during his previous service of eight years on the bench, he never rendered a decision in favor of a corporation. The Himler Coal Company rights-of-way cases were pending, and at the close of the campaign speech, I showed Judge Bandy the notes that I had made of his statement. I told him frankly that any of his adverse decisions would be appealed and that my brief never would fail to include his campaign statement.

For safety's sake, I took my rights-of-way cases to the federal courts. But during my stay in Himlerville, Judge Bandy never decided against Himler Coal Company although we had lots of unimportant law suits as a result of the mountaineers' mania for "lawing." Years later, after those poor suckers had supported their attorneys in style, their law suits were thrown out of court.

A few days after I moved to my temporary home in Warfield, I was sitting in the front room with my assistants, and we saw two men very carefully approaching our window. After they peeped into the window, one of them remarked to the other, "I told you they are whites, just like you and me." It was the natives' first look at "furriners."

I was unable to raise more money from stockholders because they were growing skeptical about the delay in the building of the bridge. But, our credit was good, and we were able to equip our very modern plant, as well as the inside of the mine, with the necessary cutting machines, motors, etc. By the time that the railroad bridge finally was ready, Himler Coal Company also was ready to load coal.

It was a holiday when the first train rolled into Martin County. I had to buy a heavy railroad engine for the two miles of railroad were operated by Himler Coal Company, or rather by a subsidiary of our company called the Kermit-Warfield Bridge Company. I borrowed two passenger cars for the occasion. The engine was manned by the timber baron, the owner of our coal land, and I was the fireman. My tears were mixed with the coal with which I fed the engine on this first trip.

My handful of loyal miners who had lived in temporary shacks for two years and some of our stockholders from nearby mines dressed in their Sunday best and were on hand to greet the first train. The guests traveling on the train, including railroad dignitaries, larger creditors, and my friends, were deeply touched when they alighted from the train and were greeted by the sounds of "My Old Kentucky Home" played on a cornet by Frank Demjen, one of the Hungarian miners.

Hundreds of Martin County natives were present to take a look at the first train in their county, and a sick old lady was carried from a distance of several miles for she wanted to see a train before leaving the wonders of Earthly civilization behind. It was a red-letter day in the history of the Himler Coal Company and in the history of my life. Since the mine was in operation and the railroad ready to haul material, I started to build Himlerville, the mining town to model after Roebling, New Jersey. Four and five-room cottages were built, each with a bathroom and all of the modern conveniences. But before building the town, we had to build the town site in the very narrow valley of Buck Creek by cutting streets on the hillsides. Water storage tanks had to be put up on the highest hill for pressure, strong pumps also had to force the water up the hill, and sewers had to be laid. These were things never before seen in Martin County.

We built about eighty dwelling houses, a large and very modern brick building in the center for offices and commissary, a public hall large enough to seat one thousand people, and a mansion for the general manager on top of the mountain. I knew that I would have to take care of visiting coal buyers, and my house was planned accordingly. The house contained sitting rooms, one of them very large for a special purpose, a den, and six bedrooms besides the servants' quarters. This house was a far cry indeed from the very modest homes of Mátraverebély and from the hobo jungles. I also encouraged individuals to build their own houses, and about thirty of them took advantage of the offer, an impossibility until then in the southern coal fields. Of course, Himler Coal Company furnished sewer, water, and electricity to the privately-owned homes. All of this sounds and seems to me now like a fairy tale but, while doing all of this, I did not sleep on a bed of roses. Just when we were ready to load coal, the coal market was shot to pieces.

Himler Coal Company did have a fair contract for the first year's output with the Norfolk & Western Railway. But, we could not load much coal before developing the main entries for more working places, and we did not make enough profit to take care of current investments. On the other hand, the collapse of the market had its bad effects on the credit situation, and I had a hard time financing the building of the town.

Banks withdrew credits from coal operators, and other creditors became panicky. But I was hopeful that I would weather the storm even though I knew that our Norfolk & Western contracts would not be renewed at the same price because coal was selling on the free market with a loss at about half our current contract price.

I knew that I had a good mine and the most modern equipment. I had a great asset over and above all operators because I personally knew thousands of miners and hundreds of stockholders and was able to pick the pick of them, the very best, who were ready to stand competition.

Few Americans know what happened to the coal industry after the First World War. Prior to that war, Americans consumed about 400,000,000 tons of coal a year but, under the pressure of war's necessity, mines were developed to produce 700,000,000 tons, almost twice as much as our country needed. On top of that, high coal prices during and immediately following World War I forced development of stockers, a more efficient use of coal by consumers, and brought the additional blessing of the introduction of fuel oil. Coal consumption fell to 320,000,000 tons, less than half of the capacity of the coal industry.

Coal operators were selling coal at a loss, bidding desperately against one another, and trying to remain in the game. Each operator hoped that enough of his competitors would be forced out before his turn came and that in future years, he would regain his losses.

Uninformed readers may ask, "Why operate with a loss?" "Why not close down and wait for better days?"

Closing down a mine of about the Himlerville size which was loading 2,000 tons of coal a day in the third year meant an expense of $40,000.00 yearly for maintenance of the mine (pumping water, ventilating, etc.), taxes, and insurance. Operators loading 500,000 tons a year were better off losing about $.05 on a ton of coal and continuing to furnish bread—even if without butter—for their miners.

The desperate coal race went on for many years. Wages of miners were reduced until some coal operators paid as little as $1.50 when factory workers were getting $5.00–$6.00 per day. But, operators still sold their coal at a loss. Over 300,000 coal miners were forced out of the only industry they knew, and over 9,000 out of a total of 14,000 coal corporations went bankrupt, with such giants as the Rockefeller-controlled Consolidation Coal Company, the Reading Company, etc., losing several billions of dollars. Few Americans know that in those days of ceiling prices, coal was the only commodity with a minimum price set by law under the Guffey Act. The Guffey Act ensured that coal operators could not sell their coal under the minimum price. Following tremendous losses and the total disintegration of the basic coal industry — alas, after the collapse of the Himler Coal Company — Franklin D. Roosevelt forced Congress to pass the so-called Guffey Act and save the remnants of the mines.

While building the town of Himlerville, I was trying to float a bond issue. Money was plentiful in our country, and the United States was buying doubtful securities from every nation in the world. But the coal industry was taboo amongst bankers. Fly-by-night brokers were willing to float coal bond issues at outrageous discounts and with very high interest rates. In some cases, discounts amounted to 25 percent, and interest rates were at 8 percent. Finally, some friends connected me with an old brokerage house in Cincinnati. That brokerage house was willing to float a bond issue of $450,000.00 on decent terms. The deal was to be consummated on September 2, 1921, my birthday. When I went to the brokerage house offices about 11:00 A.M. on that day, I was assured that papers would be ready at 2:00 P.M. and money forthcoming.

I was thus in a happy frame of mind, and I treated myself to a badly-needed suit of clothes as a birthday present to myself. My happy frame of mind ended when I was told at 2:00 P.M. that news had come from Europe about the French occupation of the Ruhr,[3] the money market had become excited and jittery, and under these circumstances, the deal was off. The bottom fell out from under me. My creditors were pressing me, and I did not see any other choice but suicide.

While in this saddened frame of mind, I was waiting for a train to take me home. I glanced at a newspaper and saw these lines: "Today is the tomorrow we were worried so much about yesterday." And I realized that there would be other tomorrows.

Instead of returning to Himlerville, I went to Columbus, Ohio, to see our company attorney who also was interested in the selling agency from whom we purchased the Ajax mine. Our attorney was one of the leading corporate attorneys in the city, and I told him that I had about reached the end of my rope. Following his advice, I called a creditors' meeting in Huntington, West Virginia. I planned to tell them that I was ready to quit.

The night before the meeting, our attorney told me that the approximately $400,000.00 indebtedness could be taken care of at a cost of $40,000.00, the usual 10 percent discount for which more conservative bankers were floating bond issues for industrial concerns. If I were willing to assure our company's attorney that I could pay back this modest amount, our attorney also would loan me $20,000.00 cash. Of course, I was willing, and the papers for the deal were drawn up immediately to that effect.

Then our attorney told me his plan of financing. I realized too late that the $40,000.00 that I had promised to pay, in fact obligated my company to pay, was the amount not for financing, but for legal advice. But the deal was closed, and I had no choice but to abide by it and realize the sad fact that I still had a lot to learn about business methods.

Approximately thirty creditors were present at our Himler Coal Company's creditors' meeting, and they presented claims varying between $3,000.00 and $130,000.00. I opened the meeting by telling them that I was helpless, unable to raise money, and ready to hand over the mine to them.

I answered lots of questions. The creditors did not seem to blame me at all, but none of them wanted to touch a coal mine with a twenty-five foot pole. It would have been helpless to auction off the mine. When the creditors were properly scared, I sprang on them the scheme for which I was to pay $40,000.00, I offered individual notes carrying 6 percent interest due in one year to be covered by a blanket first mortgage, and I requested further credits for necessary supplies on open accounts. Noting the joy with which the solution was greeted, I went one better.

Realizing that the $20,000.00 promised by the attorney would be insufficient to carry on, I told the creditors that I couldn't return to work for them with an eased mind unless I could be given a little cash for the next payroll, a matter of $8,000.00 cash. Without a minute's hesitation, the $8,000.00 was contributed pro-rata by my creditors, and I left the meeting with their good wishes, cash, and friendship.

But I still had to learn aspects of high finance when I went up to Columbus a few days later for the $20,000.00. The attorney received me with a gloomy face. He told me that a mine in western Kentucky was to be sold by the selling agency in which he was interested and, at the last minute, the deal went bad. The $20,000.00 that our attorney promised to me was to be part of the proceeds of this sale gone bad. I realized that the promised loan was not mentioned in the agreement about his fee of $40,000.00. The promised $20,000.00 was only a promise on his part, a gentleman's agreement as it were.

The attorney knew that I had to have the $20,000.00 and finally came around to say that one solution would be for the Himler Coal Company to take over the western Kentucky white elephant. In this case, he would be willing to borrow the funds and loan me the necessary $20,000.00. His bland smile at my question as to why he couldn't borrow money without our buying his mine reminded me of the so-called coffin deals in Hungary.

There were very stringent laws against usury in the Old Country, and anything over 6 percent was considered usury, a serious criminal offense. When young sons of rich men were hard pressed for cash, they went to some well-known notorious usurers and agreed to buy, for example, a carload of coffins, paying for same with a note for the full amount. When the coffin deals stunk to high Heaven, the "shipments" consisted of grain, hay, or what have you. The "goods" immediately were sold back to the usurer for cash, usually for half the amount for which the note was written.

Looking at the smiling face of the attorney of very high standing in his city, I realized that I would have to buy the carload of coffins to get the $20,000.00. Luckily, I knew a few Hungarian labor agents who failed to invest money in our company because they themselves wanted to be coal mine managers, superintendents, or other big-shots. I did not particularly like labor agents, anyhow, and I was able to unload the white elephant mine at Nebo, Kentucky, within a few days and without any other losses except some traveling expenses.

Yes, I did learn things about finances, but I had to pay for the education. However, I had a year's time to look around for a permanent solution, and I used my time for good purpose.

When I was returning from Philadelphia from an unsuccessful trip to raise money, I met an elderly gentleman on the train. He inquired kindly about my reason for being so blue. I was embittered by the impossible demands of the money sharks whom I had visited in Philadelphia, and while generally I am not in the habit of talking to strangers, especially about my personal affairs, the gentleman's kind face induced me to unload.

I told him the story of Himlerville and answered his searching questions about my previous activities from the day I landed in America without knowing the identity of the gentleman. We had dinner together and breakfast together the next morning. I was about to say "good-bye" to him at the station of Columbus where my train was due to pull out within about an hour when he gave me one of his business cards. He asked me to stay over until the evening train and call in his office at 11:00 A.M. He was Mr. F. R. Huntington,[4] President of the Huntington National Bank of Columbus. When I visited Mr. Huntington's office, he told me that he would have the Himlerville mine investigated by an expert. If my statements were proved and a bond issue approved by his man, he would take over the bond issue of $450,000.00.

Mr. Huntington further told me that he was not searching for an opportunity to make money for his bank, although that was usually to be the primary purpose of any loan. Mr. Huntington, in my case, was deeply interested in this scheme of miners' having their own mine, a scheme which might mean an entirely new solution to the increasing strife in industries.

I met Mr. Huntington many times during the following hard years, and I learned to like and respect him very much. The man was a strange combination of a financial genius, a very able painter, and every inch a philanthropist if there ever were one.

Within a few weeks, Himler Coal Company received the money from Mr. Huntington, and we paid off our notes with the creditors. When Mr. Huntington learned of my obligation with the Columbus attorney, who also was one of the attorneys of the Huntington Bank, Mr. Huntington forced that attorney to reduce his claim to $5,000.00. The world was mine, and I was all set to give all my attention to the management of the Himler Coal Company and the mining town of Himlerville.

Mr. Huntington then made an unexpected suggestion. Having full knowledge of my personal affairs and of my struggles during the previous three years, he suggested that I take a vacation before buckling down to work. He suggested that I go to see my mother in the Old Country. I had received several months of back pay and dividends from the Miners' Home, and I was well financed for the trip. So were three of my miners who accompanied me. One miner was to bring his family to America, the other was to see his parents in Transylvania, and my best friend of all the coal miners, another bachelor of about my age, Johnny Bako, was to bring his two sisters to America.

Johnny had a very interesting career before I met him. He was one of the six children of a yearly servant of a large Hungarian landowner. Yearly servants are contracted workers on a large estate, virtual peons, who can't leave their masters before the end of the year. They live on the estate with their families and, as late as 1920, a law was passed by the Hungarian Parliament that authorized the land owner to dish out corporal punishment to members of the families of such servants when, in the landowner's judgment, the family members deserved corporal punishment. The man of the servant family worked every day, and the wife usually was obligated to also work a number of days per year. Johnny was a swine herder by the time that he was twelve years old. One of his older brothers somehow came to America, and when Johnny was sixteen years old, Johnny's brother sent Johnny a steamship ticket to America.

Johnny skipped Hungary with another boy of his age and went to Bremen, Germany, to the *Norddeutscher Lloyd*[5] to purchase passage. Twelve days later, Johnny was in Buenos Aires. Forty or forty-five years ago, Argentina was reaching out for immigrants to man the vast estates. Argentina subsidized steamship companies, and since Johnny and other poor boys like him did not know the difference, they were taken by the thousands to Argentina rather than to America. After all, they had asked for the tickets to America and to an America they were sent.

The government of America's sister republic took good care of the immigrants in those days. Immigrants in Buenos Aires were housed in an immigrant's home and distributed to land owners in need of men. Johnny was sent to a German landowner and worked there for eight years. Johnny earned far less than he would have earned in America but far more than he ever dared to dream about when herding the swine of his landlord in Hungary. Eight years later, Johnny returned to Hungary to see his parents. Johnny wished to join the other emigrants from his village who had gone to America, and since Johnny had learned the difference between North and South America, Johnny came to the United States and became a coal miner.

While my traveling companions and I waited for the steamer in New York, the companions requested me to fit them out with suitable clothing so as not to embarrass me. We spent money like drunken sailors. My companions traveled second class on the *SS Rotterdam*.[6] I went first class, but I bribed the deckhand sufficiently to bring my companions over to the first class every morning, and we spent the whole day together in the large café. We spent a week in Paris, a riotous and glorious week, such as only four young men with ample finances can manage, and another few days in Munich, including Christmas. After visiting the popular night clubs in Paris, I decided to see the Grand Opera. Johnny was insistent that I buy him a tuxedo so that he would be able to accompany me.

The other boys left Johnny and me in Vienna, but my man Friday stayed with me in Paris for another week. One of my sisters from Szombathely was invited to join us. This sister of mine was made rich on my previous trip when I returned the $40.00 she loaned me on my first visit to Hungary. She was reluctant to accept the $40.00 at the time and insisted that I only return the 200 Kronen she loaned me. But at the time of the loan, 200 Kronen was the equivalent of $40.00. I forced her to accept the money which was, at the time of my first visit, a small fortune.

My sister, Johnny, and I spent New Year's Eve in Vienna and a day later, I sent Johnny to Budapest with an important message to some of my newspaper friends. One of the leading exiles of the Bloody White Regime, the Reverend John Hock, President of the first real Parliament of Hungary, was my permanent guest at Himlerville. This was known by the authorities of Hungary. Before going into Hungary, I wanted to know whether I would be molested by the authorities. When proper inquiries about my safety had been made, Johnny Bako was to call me up by phone to give me the green light or to tell me to stay out of Hungary. He also was to reserve a room for me in the Ritz Hotel.

Next day, Johnny called me to take the train for Budapest, and he took me to the Ritz from the station. When I inquired about his room, he made a remarkable statement showing the intelligence and the fine tact of a Hungarian peasant. Johnny replied, "I took room in the Central Hotel and reserved room for you here because we can't stay together in Hungary. It was all right in Paris, in Vienna, for I am dressed just as you are, and we had to speak English. But in Hungary, the minute I open my mouth they will realize that I am a peasant, and I will constantly embarrass you."

And hard as I tried, I could not induce Johnny to move to the Ritz. But I did convince him to spend most of his time with me. I did take him to the best newspaper circles of Hungary, and he endeared himself wherever we went with his tact and his native intelligence. I introduced him to members of my family and to my mother, and they became very fond of him. With brutal frankness, I told my family that Johnny was nearer to me than any of my brothers with whom I had to be re-acquainted on each trip. I need not say that my mother, as well as all members of my family, were very happy and very proud of me. Himlerville was known in Hungary by the general public, and I was a famous man.

The village notary and some other local dignitaries of Mátraverebély called on me to pay their respects and to obtain some funds for the restoration of the church. This pleased my mother immensely. I also was happy knowing that Himlerville was financed and hoping that my success would be permanent.

I have to tell you about an interesting incident that shows more clearly the caste system of Hungary than anything that I can say about the caste system. Once I took my mother and one of my sisters to Szombathely, and we met Johnny by chance at the station of Budapest. I did not know that Johnny was to take the same train for Pápa[7] to see his sisters, and we were glad to see him. He already had his ticket, and with the remark that the second class is usually overflowing with passengers, advised me to follow his example and buy tickets for the first class. This I did, and we traveled in style with the four of us in the same compartment. This was the first trip that my mother and sister had traveled in a first-class coach.

An hour later, a gentleman joined us in the coach compartment for eight and greeted us very politely before he took his seat. Within a few minutes, Johnny said something to my mother and mentioned my name which, of course, was well known in Hungary. Our fellow traveler turned to me with the question, "I beg your pardon; I could not help hearing your name. Do I have the privilege to meet Mr. Himler of Himlerville?"

Mutual introductions followed. The man was a high official in the Foreign Office of Hungary and for about two hours, we had a very pleasant conversation touching on the politics, literature, and art of America and Hungary.

About two hours later, Johnny was looking out through the window and at one point, he said excitedly, "Here is where I used to herd the swine in my youth!" The three of us took Johnny's exclamation for granted and thought it only too natural that Johnny was happy to see his old hunting grounds. But our new friend grew rigid, as if lightning had struck our compartment.

Our new friend asked, "I beg your pardon; what did you say?"

Johnny replied, "I said that when I was fourteen years old I was a swine herder at the spot I pointed out. Why are you amazed at the statement? I am only a coal miner now, working in the mine where Mr. Himler is the manager, or director as he would be called here."

Our new friend got up and left the compartment without saying another word, not even a "Good Day." Evidently, he was offended by having been enticed by some trick into associating with a swine herder and he felt that the whole atmosphere in the compartment was contaminated.

My mother, my sister, Johnny, and I burst out laughing, but I was—and still am—mad, for I knew no finer man than my friend Johnny. If either Johnny or the Hungarian official could be called a gentleman, there is no doubt in my mind that the real gentleman was Johnny, my fine swineherder friend.

Prior to the first general election in Hungary, one of my friends and I sent a few hundred dollars to Victor Drozdy, leader of a small liberal group campaigning in the first general election in Hungary. When I arrived in Hungary, Drozdy proposed to organize a banquet for me, and I had a hard time convincing him that if such banquet were held, I would not attend. Two weeks later, he invited me to dinner at his home. I was unpleasantly surprised to see forty members of Parliament present and menu cards stating that the banquet was in my honor. Speeches were made, but I remained silent.

One of my newspaper friends, also present, expressed his surprise that Mr. Vazsonyi, the most outstanding liberal of Hungary, was absent. Later, my newspaper friend informed me that I had offended Mr. Vazsonyi, "the most outstanding liberal of Hungary," at a meeting in Parliament by calling him "Congressman" instead of "Your Grace." According to my newspaper friend, Mr. Vazsonyi was Minister for a few days during World War I, and the title of "Your Grace" was due him on this ground.

Drozdy hit me for a loan of $500.00 to finance his newspaper before I left Hungary. A few months later, Drozdy's shady dealings compelled him to emigrate to America. He published a newspaper in Chicago, became an American citizen, and lived in the United States for a number of years. When Nazism[8] became strong in Hungary, Drozdy returned there. He changed his allegiance once more, entered public life again, and profited much. In fact, he became wealthy by taking over positions and factories of Jews. My attorney approached Drozdy with my claim for $500.00, and Drozdy intimated that if I pressed the claim, my relatives in Hungary might suffer. Drozdy is still rich, active in public life, and I look for him shortly to enter the Communist Party. There are a good many of his ilk in Hungary at present.

There were two other Hungarian-American newspapermen in Budapest. Miss Ilona Fulop, who had joined Johnny and me in Budapest, and

Johnny and I spent many pleasant evenings with the newspapermen in the Café New York. We listened to any American group's music, with Johnny's demanding at least once every evening that the group play "My Old Kentucky Home."

Carrying the customary poppy seed biscuits and hearing the usual whispered last words of Mother, I left Hungary after a two-month visit. Johnny and I returned to America, to Himlerville, and I picked up the work where I had left off before my trip. The Himler Coal Company was still loading coal for the railroad at a fair price, the superintendent had managed the operation on an even keel, and I started to organize life in the camp.

CHAPTER 16

Life in Himlerville

★

I BUILT A HOUSE for the *Magyar Bányászlap*, equipped a printing shop — the first printing plant I ever had —, and moved my newspaper to Himlerville. To my regret, Miss Fulop was unwilling to leave New York so the newspaper was put into the charge of my oldest nephew, Andrew Fay Fisher. My nephew was the only relative I had in America. He joined me here in 1921, became the owner of the *Magyar Bányászlap*, and also supervised the publishing of the newspaper in Pittsburgh.

To combat the antagonism of the local Martin County population, I engaged a local man to edit the only English weekly newspaper Martin County had ever had, the *Martin County Advertiser*. The extent of this antagonism, artificially created and fed by some local politicians, was so great that for over two years, I had to carry riot insurance. I kept a deputy on the job day and night for I feared that an attack would be made on my men. Slowly, but very slowly, I won over the best of my antagonizers, mostly by giving them employment in the mine and also by treating them as well, or better, than the company's own stock-holder workers.

I was against the common practice of charging outrageous prices in the mine commissary and thus keeping out competition. I encouraged the establishment of small private stores to force the commissary managers to compete. Even antagonistic natives took advantage of our low prices. We sold everything from steak and ice cream to ladies' hats. Our best year provided us a business of over $400,000. Sometimes I went to New York to buy men's and women's clothing in job lots, and I put on sales drawing crowds even from nearby towns in West Virginia.

The Martin County situation worried me and the miners. While native coal miners seemed to care very little about schools, immigrants were unwilling to stay in any camp, no matter how well satisfied they were with the work, if schools were not provided for their children. There is nothing surprising about this for immigrants feel their own handicaps, and they wish to save their children from the same fate. It may be due to this fact that during World War I, the percentage of non-commissioned officers with foreign-sounding names was so high. Of course, Martin County was obliged to provide for the immigrants the same four months of school that the local children had, but no Hungarian parents would remain at Himlerville and put up with such an arrangement for the education of their children.

I asked the Kentucky Legislature to create the town of Himlerville, and then the town council took matters into its own hands. No, I was not the mayor of the town. I bestowed that honor on my secretary, Eugene Lang, who was also secretary and treasurer of Himler Coal Company. We taxed ourselves and individual property owners within the town limit over $4.00 per $100.00, built a modern school house to take care of two hundred children, and I applied to the Kentucky Board of Education for eight of the best available teachers. I also offered salaries between $200.00 to $225.00 against the $40.00 the teachers of Martin County then received. Two years later, I received a letter from the Kentucky Board of Education, and the letter informed me that Himlerville had the best grade school in the whole state of Kentucky. Following the experience of the First War, when twenty thousand illiterate draftees of Kentucky had to be taught to read and write in moonlight schools, I thought of those draftees' leaving their parents in the future. I knew that the parents would never learn to read and write English. Hungarian, therefore, was included in the Himlerville School curriculum, and it was compulsory. Children of our native workers, even those living outside of the town limits, were permitted to enroll in the Himlerville School, and we enrolled about thirty of them. I had quite a kick when, on one of my birthdays, I received a Hungarian letter from each of the native students to thank me for giving them the opportunity to be enriched by knowledge of a foreign language.

I received some education about public utilities and holding companies by the taxing of the properties within the Himlerville limits. Three gas wells were included in the town, and, of course, Himlerville assessed them for city tax. The gas company objected, appealed, and ran to the Public Service Commission, but finally the gas company was forced to pay the assessed tax. As a result of this fight, I discovered that while customers right at the source in Martin County were fleeced $.40 for the gas, consumers one hundred and thirty miles away in Ashland, Kentucky, paid only $.26. The gas fight also gave me the pleasure of meeting four different gas companies on top of each other, each raking off some profit, each having a finger in the pie.

A small Catholic church was built in Himlerville, and a priest was sent there by the Bishop of Covington, Kentucky. However, on important holidays, a Hungarian priest was invited from one of the large Hungarian colonies. I extended the same services for my Protestant families by engaging a Protestant minister as the Hungarian teacher in the public school, and he held Protestant services in the public hall.

With the consent of the Himlerville working men, $1.00 was deducted from each Hungarian worker, and with the necessary contribution by the Himler Coal Company, a brass band of twenty-four members was organized. An ex-bandmaster of the Austro-Hungarian Army was engaged to teach and conduct the band. The band furnished surprisingly good music within a few months. Amateur shows and dances were held every week with no other charge for admission except the $1.00 we deducted monthly from the men. Moving pictures were shown twice a week and were attended by many of the native population as well as by the population of the camp.

I conducted a night school to teach civics and to prepare Hungarians for American citizenship, and the night school was greatly appreciated. It was said, within a few months, that Himlerville was manufacturing American citizens. Miners often secured employment for the sole purpose of attending the night school and obtaining citizenship. I usually took twenty-five to thirty prospective citizens to the federal court of Catlettsburg, Kentucky, where citizenship hearings were held three times a year.

Once I took about twenty-five prospective citizens to the circuit court at Inez. Instead of the examiner's coming down from Cincinnati to conduct the proceedings, the examiner deputized me to conduct the citizenship hearing at Inez, and I made quite a show of it. Inez never had been the site of a citizenship hearing, and the courtroom was full of curious natives. Instead of asking two or three questions of each prospective citizen, I threw the book at each one and never asked fewer than fifteen to twenty questions about the Constitution of America.

The audience was dumbfounded at the knowledge of the Hunkies, a knowledge which was far above what all the natives knew about our Constitution. I derived much satisfaction when an old native fellow made the audible remark that, "We took the country from the Indians because we were smarter, and we deserve to have the country taken over by these people because they are smarter than we are."

Himlerville held the balance of political power in Martin County within two years. Politicians had to consult me in nominating a sheriff, and U.S. Republican Congressman John W. Langley, as well as Mrs. Katherine Langley after her husband was convicted of illegal alcohol sales to New York bootleggers during prohibition, were elected by four hundred votes of Himlerville out of a total of eight hundred votes in Martin County.[1]

Himler Coal Company's demonstrated goodwill to all and the influence of the *Martin County Advertiser* changed the attitude of the local people, the number of our friends increased, and most of our antagonists became at least neutral. The style of life in Himlerville influenced the whole of Martin County. Log shanties were repaired and many were equipped with windows. Natives earning good money in the mine built themselves decent frame houses, their diets were changed for the better, and especially, the native women started to dress better. Some of Himlerville's neighbors applied for electricity, and Himler Coal Company never turned down any reasonable request to improve the standard of the community.

When Christmas time came to Himlerville, presents were placed under the tremendous Christmas tree in front of my house and then given to hundreds of children whose parents had nothing to do with Himler Coal Company. Two years later, several native families had their own Christmas trees for the first time in their lives and for the first time in the history of the backward Martin County.

Some of the native politicians did not like the growth of the Hungarian influence, and during the heyday of the Ku Klux Klan, a local Klan branch was organized in Kermit, West Virginia, for the expressed purpose of chasing out the foreigners from Martin County. Without my asking them, some of my loyal native workers joined the Klan branch and kept me posted on developments.

These loyal workers and Klan members from Martin County had to wade the mid-winter, knee-deep mud to attend the Klan meetings on the

West Virginia side. I called the leader of the Kermit Klan to my office, and I told him that I knew his purpose and did not wish to interfere with his Klan in any shape or form. However, in line with the policy of the Himler Coal Company, we still wished to be of service to our community, and thus I then offered the Klan leader the free use of our company's hall to save his members from the muddy trip to Kermit. I had the same offer published in the *Martin County Advertiser*, and the Ku Klux Klan was laughed out of existence.

Traveling salesmen told the Klan tale in Huntington, West Virginia, a city of about sixty thousand. A committee of Huntington Jewish merchants called at Himlerville and requested my assistance to combat the strong Klan of Huntington. I then organized a local Huntington committee of Catholic priests and Jewish leaders, and we called on the owner of the largest Huntington department store. I requested the owner to state whether he was a member of the Klan or not, and I told him frankly that we would organize a boycott unless he came clean.

The department store owner indeed was a Klan member, and he threatened me with consequences by pointing out the law against boycott. I, on the other hand, reminded him that the tenets of the Klan mean boycotts against foreigners, Negroes, Catholics, and Jews, and we thus were going to defend ourselves with the same weapon. I had some standing in the Huntington community because the Himlerville people did a good bit of trading in Huntington. The Huntington priests also were ready to join me in a campaign against the Klan. The aforementioned merchant, as well as four other businessmen of high standing, were forced to publicly announce their withdrawal from the Klan in the local Huntington daily newspaper.

To this day, I do not understand the attitude of the foreigners, Negroes, Catholics, and Jews during those days. I can't see their reason for often taking the abuses of the Klan lying down instead of fighting the masked terrorists openly.

The most interesting part in the existence of Himlerville was the reversed social condition regarding the relationship between native and immigrant citizens. Everywhere else in America, the immigrant group was the under-privileged, and it was hard for any of them to crash the gate of the so-called 100 percent society. However, in the very poor Martin County, the immigrants made more money and maintained a much higher standard of living than even the leading native men in the community, such as attorneys and merchants. If there were any uplifting done, it was done by the immigrants.

As the enmities and antagonism between Martin County natives and immigrants ceased, native physicians, attorneys, teachers, as well as farmers and coal miners considered an invitation to the social gatherings of Himlerville a distinction and readily availed themselves of the privilege of attending. The height of social standing in Martin County was reached by a young girl or boy if she or he were invited to participate in the Himlerville amateur show or to play a part and appear before the audience composed of "furriners." Amateur shows always were performed in half English, half Hungarian.

When there were cases of confinements or illness, the wives of the Hunkies usually hired a poor native woman to take care of the house. However, in other localities, immigrant women were doing housework for the native families.

While, of course, the main objective of the Himler Coal Company was to operate the mine with a profit, my personal objective was to raise the standard of my people in every respect. My people were encouraged to live up to the standard in their modern and much-appreciated homes, and visiting Americans were astounded to see coal miners eating off white table cloths and using white napkins.

In addition to teaching civics to the immigrants, lectures were held about America, the history of this country, and every aspect of American life. Engineers, office workers, and teachers were glad to alternate with me as speakers and teachers. Our native friends were invited to lectures about Hungary, the Hungarians, and about the rest of Europe and Europeans, and appreciation of the lectures was demonstrated by an ever-growing attendance.

Himlerville became known all over America and the world as an interesting social experiment. Magazines like the *Survey Graphic*[2] published illustrated articles, the *Chicago Daily News* sent a man down to write a series of articles, and editorials were published as far away as Dallas, Texas. Only the Hungarian leaders remained unreconciled. Only the radical Hungarian press was opposed to Himlerville and predicted gloatingly that the town would eventually fail. One of my miners told me once that life on the camp would have been a paradise were it not for the mine. And he was right.

The Himler Coal Company was nominally managed by a board of directors composed of nine coal miners, my assistant, the secretary-treasurer, and myself. Directors were chosen at the annual stockholders' meeting which was a holiday in Himlerville. The stockholders' meeting always was attended by four to five hundred coal miners from far and near. The large attendance was not so much an occasion to interfere with the management of the mine as it was an opportunity for the stockholders to have a joyful reunion with friends and to investigate the possibility of their moving to Himlerville with their families. The Himler Coal Company had close to two thousand stockholders and could not employ over three hundred men. The company always was under pressure by some stockholders.

The average individual stock investment in Himler Coal Company was a little less than $300.00, with some stockholders investing from $2,000.00 to $3,000.00. Many stockholders invested only $100.00, not for any hope of gaining much profit on the small investment, but to further the scheme of having at least one permanent home of the Hungarians in which they could take pride. Stockholders were very proud of Himlerville and with good reason. Coal operators acknowledged that it was one of the best laid-out, best-equipped mines in the district with a modern and model camp unparalleled then on the southern coal fields.

I can't say that Himler Coal Company had two thousand angels for stockholders. About 95 percent of them were the cream of the Earth, but there were a few expecting easy jobs, privileges, and special considerations in assignments of jobs because of their $100.00 – $200.00 investments. I had to be firm to weed out the trouble makers, and I encouraged those with quarrelsome wives to move or I assigned them a house far from the others.

I was accused at times of being an iron-handed dictator, but I never failed to listen to just complaints. Complaints were always in the presence of the superintendent or the foreman against whom the complaint was made. I listened to and I adjusted every just grievance. The Himler Coal Company Board of Directors never had to criticize any of my actions, nor did my stockholders.

I took pains to educate the members of the board to the task of managing a million-dollar corporation, and I gained much technical knowledge from the experienced coal miners. I also visited the Himlerville mine and the men at their working places several times a week. The harmonious relationship between my men and myself was greatly helped by the fact that my door was always open to any of my men. There were very few evenings when my larger living room, planned for the purpose of large gatherings, did not have fifteen to twenty miners for an hour or two. The miners there were not always the same, excepting my buddy, Johnny Bako, who was one of the best coal loaders but who spent most of his free time in my garden or in my house. The miners and I talked about the mine, about the camp, about politics, and without the constant companionship of those men, it would have been hard for me to live on a mining camp. Sometimes four or five of them retired to my den to play cards, and they used my home as a sort of club, even if I were absent from the place.

Visiting men coming to Himler Coal Company stockholders' meetings were put up by the families of the camp. The families fully informed the visitors about the management of the mine, the pleasant life on the camp, and my close relationship with my own men. Although the company had very hard goings after the first year and ever increasing difficulties until finally we had to go broke and bankrupt, no directors' or stockholders' meeting ever had one word of criticism about the management of the company. The stockholders' confidence in me or my good faith never wavered.

Stockholders' meetings usually closed with a picnic, and hundreds of men went back to their respective camps to tell the other stockholders that their money was invested honestly and with proper planning. The stockholders told their co-workers that the use of their money was visible at Himlerville and that the co-workers could disregard the libelous and malicious ravings of the radicals and other enemies of Martin Himler or Himlerville.

I had a fair salary, somewhat less than the usual scale for the position that I held. The stockholders at each meeting tried to raise my salary, even when we were operating with tremendous losses. The stockholders knew that I had to maintain a large house to entertain coal buyers. I was the largest Himler Coal Company stockholder because I invested all of my $10,000.00 in the company, and stockholders' meetings repeatedly offered me bonuses in stock for promoting the company and for advertising the sale of the company's stock. The *Magyar Bányászlap* did not accept any money for the advertising and promotion of the Himler Coal Company. Nor did the company ever order any printing in my printing shop, although my shop did commercial work for neighboring companies.

The Himler Coal Company had to fall in line with the coal market and lost money in the second year of our operation. Just as our competitors did, the company tried to remain in the coal race by cutting wages and by economizing everywhere, and our miners—well aware of the situation elsewhere—always accepted the new sacrifices gracefully. The Himler Coal Company Board of Directors asked our miners whether the company should accept a large coal order and assure the miners of steady employment with lower wages, or turn down the order and try to remain with our current output on the open market. The miners knew the price that the company received for every ton of coal, and detailed cost sheets were posted every day. Our miners chose to accept reduced wages with steady employment. Of course, I took good care to have the *Magyar Bányászlap* publish all the sad facts about the coal mining industry and about the failures of hundreds of other older and financially stronger mining companies. Those facts were substantiated by hundreds of other companies' miners coming to us to look for work when the mines where they were employed were shut down.[3]

Mr. Huntington stood by the Himler Coal Company during three years of steady losses. He financed our losses and encouraged me to carry on in the belief that the whole coal mining industry couldn't go bankrupt and that eventual consolidations would bring about relief. I grew old during those unhappy years, and I was forever aware of my responsibility to Mr. Huntington, to my stockholders, and to my people at Himlerville. When attending the Himlerville Saturday evening shows and dances, I was invariably thinking of my duty to provide bread and butter for my people. I did not share the optimism of Mr. Huntington.

The pressure of the creditors became unbearable, and I begged Mr. Huntington to put a stop to the agonies. But, he pointed out to me that he did not wish to lose hundreds of thousands of dollars, that he had invested the money with faith in me, and that it was our common duty to carry on as long as possible. To relieve me of the pressure of creditors, he agreed to voluntary receivership of the Himler Coal Company. One of the directors of Mr. Huntington's bank, Mr. F. G. Sutton, was appointed receiver.

I was lucky to always have good friends in my life and to meet kindness, but never did I have a better friend and more kindly co-worker than Mr. Sutton. As a result of his applying his vast experience as a coal operator and giving preference by selling Himler Coal Company coal before he sold his own, our company reduced our losses to the minimum. Of course, the receivership was greeted by joyful jeers in some Hungarian circles, particularly by the radicals, but my miners and stockholders were greatly and pleasantly surprised that we did not go completely under.

Then the final blow fell.

An inch of slate appeared in the middle of the five feet of Himlerville coal, and the thickness of the parting grew by each day until it reached nine feet. Of course, I did not worry much at the beginning. We had put down core holes at every two thousand feet in each direction before leasing the coal, and we knew that eventually the slate would disappear. But we did not expect, nor were we able to cope with nine feet of slate. Since the slate cut the coal in the middle, it was impossible to mine on top of the slate, or to mine under it, and we had to mine nine feet of slate with five feet of coal at a time when even clean coal was mined at a loss. And, to boot, we had to carry most of the slate out of the mine at an additional cost.

The excellent sandstone roof of the Himlerville mine also disappeared about the same time, and we had to start supporting the roof with timbers at a cost of about $.20 a ton. Coal was then sold at $1.25 per ton, and every additional penny was an unbearable burden. Hoping that eventually we would leave the slate behind and at the same time regain our good roof, we carried on for two years. Mr. Huntington was always ready to meet losses and always encouraged us to keep on, although the receiver also was indebted to close to $200,000.00.

As a form of intellectual recreation for myself, I started to publish a small Hungarian weekly newspaper called the *Martin Himler Weekly*. I also carried on a fight in the *Magyar Bányászlap*, a fight of which I am still proud. The small mining towns of the southern coal fields offered all sorts of recreations to coal miners. Not the least amongst the recreations were the various dives and brothels, and venereal diseases were rampant amongst miners. Businessmen of those small mining towns were encouraging the existence of the numerous brothels as incentives to tempt the miners into the town to spend their money.

I started to demand a general cleanup of the brothels. I knew that lots of single men would always patronize and would, in fact, be in somewhat of a need for the services of the oldest profession. But, I maintained that physical examinations should be provided and that brothels should be cleaned of the bums, bootleggers, and other scums who fleeced the coal miners.

The authorities did not take kindly to my articles (published in English and Hungarian), but the public did support me. The fight went on for several weeks.

The local West Virginia newspapers remained silent for a time. But, when I started to publish the names of owners of real estate that housed brothels, the best and most powerful daily newspaper of the field, *The Huntington Herald-Dispatch*, joined my fight. Then, the Charleston and Logan newspapers had to fall in line. I was proud to have my foreign language weekly lead a fight joined by several much larger English newspapers and to have the West Virginia State Public Health Department endorse my fight. I took somewhat less pride, but much amusement, in the endorsement of my Hunky newspaper and myself by the Ku Klux Klan of Logan, West Virginia. Our fight did result in somewhat of a clean-up and in required medical examinations of the prostitutes of the field.

Once I was sick with a touch of pneumonia and while the excellent Hungarian physician was sitting at my bedside, I told him that I felt a strange dry spot, about the size of a dime, on the roof of my mouth. The physician looked into my mouth, and I noticed that he got very scared. And upon my insistence, he told me that the spot might be cancer or syphilis. Nice choice, anyway you take it, and the prospects were in line with my mood in those desperate days.

CHAPTER 17

The Lost Dream

★

NEXT DAY, I went to Columbus, Ohio. I told Mr. Huntington about the fine chances I was facing and begged him to put a stop to our miseries, take the mine off the hands of the Himler Coal Company, and relieve me of my obligation to him. Mr. Huntington felt sorry for me and, after consultation with Mr. Sutton, agreed to wash his hands of the mine.

I called on Dr. Crotty, a famous cancer specialist in Columbus, and he cut off the spot in my mouth for a laboratory examination. He also took blood for the various syphilis tests. And he tried to assure me—in good faith, I believe—that even in the case of cancer, it was in its earliest stage and could be arrested. I remained in Columbus, awaiting the results of the tests. All results were negative but, of course, I could not be sure. If the truth has to be told, I did not much care. I was about to face my people to tell them that everything was over, that our beloved town and mine were wiped out.

Mr. Sutton visited Himlerville very often during his two years of receivership. He had met all of my men, met lots of them several times in my home, and he knew the existing friendship between the men and myself. But, he did not know them well enough. Mr. Sutton was fearful that the miners would attack me when they learned that the Himlerville mine was closing. Mr. Sutton offered me $2,000.00 and advised me to go to Canada before he announced the end of Himlerville.

Next day, I went home. But, Mr. Sutton was still fearful and came down to the mine with me. He was deeply touched at the final meeting that I called to announce the end, and my tearful people offered me sympathy instead of abuse. They knew well that I had lost the most amongst them. The following day, many of us left with the same train. I was on my way to the Mayo Clinic, and my miners were on their way to look for homes elsewhere in cities or other mining camps. I said good-bye to every person in Himlerville before I left, and I never returned to Himlerville.

The properties of the Himler Coal Company, costing about $1,250,000.00, were sold for $50,000.00 and the purchaser—failed. The final Himlerville bankruptcy hearing gave me a small satisfaction when I heard from the lips of the Judge of the Federal Court of Columbus that Himlerville was one of the cleanest bankruptcy cases in his court, and that Martin Himler, the worst-hit victim of the circumstances, reached his end with clean hands. I had only one satisfaction with the tragic end of Himlerville, although in later years, I realized that the decision from which I derived this satisfaction was perhaps the greatest mistake of my life. I never ceased to love Icza and to this year, I have a tender feeling for her. The feelings intensified perhaps because I found her in sad circumstances in 1945 when both of us were old. But all wounds heal slowly. I, just as any other healthy young man, longed to have a family, especially when I was alone in my big house in Himlerville.

Although my family was very poor, we were a happy lot and, in my youth, I always hoped to have lots of children. The house at Himlerville was planned to take care of a sizable family and during the two years of building, I was waiting to finish my work and then marry.

I was very fond of a Hungarian girl in New York, and though we were not in what one would call flaming love with each other, and no definite mention was ever made of marriage, there was a sort of silent understanding that eventually we would drift together. Himlerville was ready, but I could not decide to establish a family while my finances were shaky. The girl did not know my reason for the delay. She married someone else.

A year later, I met another girl whom I liked very much and with whom I think that I was in love. I was on the verge of marrying her when I was scared off by the marriage of my assistant. He brought a girl from New York to Himlerville and within two months, there were complications. His bride came with exaggerated ideas about the coal barons, and she demanded luxuries beyond reason. As a matter of fact, her extravagance drove her husband to suicide within eighteen months.

Being pessimistic by nature, I never shared my banker's optimism of finding a solution to the ills of the coal mining industry in time to save Himlerville. I knew that eventually I would leave Himlerville bare-footed. I did not feel that I could bring a wife to live in a mansion with several servants only to force her within a short time to share my exile and a very uncertain future. I felt at times that I never could part with my life's work alive. Other times, I felt that my prospective collapse would unfit me for the company of sane and decent people.

To this day, I do not know whether it was decency or cowardice that prevented me from marrying, though I suspect it was the latter. This frustration embittered me even more than the loss of my dreams and, in later days, I spent very many unhappy evenings in my bachelor home. But when the Himler Coal Company went broke, I was glad that I stuck to my decision to not bring a wife to Himlerville.

I arrived in Rochester, Minnesota, with $4.00 in my pocket. That was all the cash that I had at my disposal. I expected to be put in a hospital where shelter and food would be provided for a few days until my nephew—enroute then from Hungary—arrived. My nephew, Andrew Fay Fisher, was the beneficiary of some of my life insurance policies, and I needed his signature to borrow money.

I was examined at the Mayo Clinic and told to report every third day for treatment. No hospitalization was offered. I went to a tourist camp and hired a cabin for $.75 a day. I took my meals in a lunch room and hoped that my nephew would arrive in time.

I had announced the final collapse of the Himler Coal Company in the *Magyar Bányászlap* before I left Himlerville, and in a separate notice, I begged my friends who owed me money to send money orders to Rochester. I told them about my desperate plight. I had loaned approximately $6,000.00 to various friends and, within three days of my arrival in Rochester, I had plenty of money. However, the money didn't come from all those indebted to me. Miss Fulop had just sold a story called *Spring Shower* to be filmed, and she mailed me $100.00. Other good friends, who did not owe me, sent me checks amounting to over $400.00. Twenty of my miners, most of whom lost their investments in the Himler Coal Company, sent me amounts varying form $5.00–$25.00. The $6,000.00 owed to me was paid to me in sums of $10.00 and $70.00 from two coal miners. The outstanding money, loaned to Hungarian intellectuals, never was paid.

Mayo Clinic told me within eight days that there were no malignant tendencies in my wound, and after I received three X-ray treatments, I was sent—home? Cured. But I had no home to go to.

I returned to Columbus, Ohio, to attend the bankruptcy proceedings, and several miners came to me with the proposition of buying in again and promising to put up the necessary capital. While I was in Rochester, Martin County was visited by an unprecedented rainfall. The usually dry Buck Creek was on a rampage and about twelve houses were carried away. I took this to be a notice served by God Himself that Himlerville was wiped out, and I refused to consider the miners' offer. Our successor's failure—with an investment of only $50,000.00—proved that I was right.

I endorsed notes for the Himler Coal Company for over $100,000.00, and I had to plead bankruptcy. I lost all of my personal property except the *Magyar Bányászlap* which was a separate corporation.

Mr. Huntington inquired about my personal plans, and I told him that all I had left was the printing shop at Himlerville and my newspaper. I told Mr. Huntington that the newspaper would very likely be deserted by its subscribers, thousands of whom lost money at Himlerville and, perhaps, also lost faith in me. Mr. Huntington advised me to move to Columbus, and he offered to secure credits for me with the local newspaper houses. He pointed out that, at worst, I could try to make a living with my printing shop. I borrowed money on my insurance policies, moved my shop to Columbus, and rented a furnished house with my nephew, Andrew Fay Fisher, and a young Hungarian chap who had been postmaster at Himlerville and who had decided to throw in his lot with my nephew and me.

During the final Himlerville bankruptcy hearing, I requested that my good friend, Congresswoman Langley, intervene with the Kentucky Postmaster General and change the name of the Himlerville Post Office to eradicate the memory of my failure. Mrs. Langley was very fond of the Himlerville camp, the nicest place in Martin County with its well-cared-for buildings and beautiful flower gardens. Congresswoman Langley selected the new name of Beauty, Kentucky, to replace Himerville, and Beauty is now the name of the lost dream.

The final collapse of Himlerville was an earthquake-like sensation in Hungarian America. The radicals, as well as every Hungarian leader who was ever offended by me or by my newspaper, and other individuals who were just jealous, made the most of the collapse. They had a Roman holiday. Most of the Hungarian newspapers, but especially the Communist and

Socialist publications, carried on for weeks and denounced me as an adventurer and a crook. The kinder publications called me swollen-headed and said that I was a stupid dreamer. A few Hungarian newspapers came to my defense by pointing out the total collapse of the whole coal mining industry and reminding the miners that their money was invested in good faith.

Nothing except the attacks by the antagonistic press was published in the *Magyar Bányászlap* for five weeks. I published every article, every charge, without omitting one word and without publishing the friendly remarks or defenses of any sort. I wanted to know how I stood with my miners.

During the fifth week following the collapse of Himlerville, I received hundreds of letters from miners who requested me to stop pouring salt into our common wounds. Many sent payment for two to three year subscriptions of *Magyar Bányászlap* to help me in my own difficulties and to demonstrate their loyalty. I also received a cable from the then William Randolph Hearst of Hungary, Nicholas Andor. Andor owned three large daily newspapers in Budapest, and he offered me the position of Chief Editor with one of his newspapers. The offer was followed by a letter stating that I could state my own terms and that I would have a free hand with the newspapers. I appreciated the offer deeply, but I did not hesitate one minute about turning it down.

I felt that in Hungary, I would be just as green as I was in America when I arrived here in 1907, except that I would know the language of Hungary. I could not hope to beat the excellent Hungarian newspapermen at their own game on their own ground. I knew that if I ever were able to regain my self-respect and the respect of others, I would have to stage my comeback right here in America. I did not ever have any desire to leave America, my country.

A few weeks later, the New York banker with whom I was in Hungary in 1918-19 arrived home from a visit to the Old Country. He brought me another message from Andor, a message imploring me to change my decision.

A few years later, I met Andor, and I asked him for his reasons for trying to entrust his newspaper to a green man. I received a very logical answer. Andor realized that America was offering mankind something better than Europe was giving, and he wanted to publish a daily newspaper edited by an American and written from the American point of view.

The small Hungarian colony of Columbus, Ohio, received me with understanding and kindness for which I forever will be grateful. As a hobby in Himlerville, I had published a small weekly newspaper called the *Martin Himler Weekly* for intellectual Hungarians, and in Columbus, I also published a small local Hungarian weekly in addition to the *Magyar Bányászlap* and the *Martin Himler Weekly*. Including the three newspapers and some income from commercial printing, my nephew, my postmaster, and I were able to make both ends meet. Each of the three of us drew a salary of $30.00 per week, just enough on which to exist. I had little desire for anything else. My faithful Hungarian cook came up from Himlerville, kept house for us, and did everything to provide us with a tolerable home for a very small wage.

Mr. W.H. Warner of Cleveland, Ohio, who once bought large quantities of coal from the Himler Coal Company, looked me up shortly after our settling in Columbus. He extended the usual invitation for the opera season of Cleveland during which I used to be his guest. Mr. Warner was astonished when I smilingly told him that my income of $30.00 a week did not include expenses for such luxuries. Two weeks later, I received an offer for the job of personnel manager and welfare man at Mr. Warner's more than twelve coal mining operations. A salary of $5,000.00 was included, but I did not care to see another coal mine, and I declined the offer with heartfelt thanks.

My nephew, my postmaster, my cook, and I were organized within a few weeks, and we adjusted our lives to the new circumstances. I wrote the three newspapers, my nephew administered the business end, and my other young associate solicited ads and subscriptions for our new local newspaper. I soon realized that instead of losing my miners, our common misfortune had become a new and even stronger bond to tie us together. The misfortune had developed a sort of a freemasonry between the victims.

Indeed, a few years later, and upon the request of the miners, we started to organize a get-together, something like a home-coming day of the ex-Himlerville residents. We met every two or three years, sometimes on the southern coal fields and sometimes in larger cities where many of the miners had found homes after their exile from Himlerville. I derived immense joy in meeting not only the old folks, but also and especially their children, most of whom appeared at these gatherings with children of their own. And if a lone bachelor ever had a family, I certainly had a large one, for I, as well as they, considered ourselves one big family.

I felt the greatest satisfaction and the deepest appreciation at the first of such homecomings held at New Camp, Kentucky, about twenty miles from Beauty. A committee of Martin County natives called on me and requested me to return to the old place and put life into the old operation once more. The committee offered the first $5,000.00 to the necessary capital and assured me that I would be welcomed by the whole county. The spokesman of the committee was John Pinson, the son of the Ben Pinson who was the first to sue the Himler Coal Company and to threaten me with a shotgun when we crossed his land with the railroad.

Noting the disintegration of the coal industry and knowing that hundreds of thousands of miners eventually would have to leave the industry, the *Magyar Bányászlap* carried on a campaign to lead the Hungarians back to their age-old occupation—farming. I approached railroad companies, timber companies, and oil companies with vast tracts of land, and I induced several of them to organize Hungarians. Several small Hungarian farming colonies were established by various interests in the southern and northwestern states.

I had some correspondence with the Chicago, Milwaukee, St. Paul, and Pacific Railroad before I left Himlerville. Once, I met their Commissioner of Agriculture who was sold on my Hungarian farming idea. It happened, though, that the railroad's vice-president was not in favor of the agriculture colonies, and "protocol" prevented the commissioner from going over the head of the vice-president to the president. The commissioner came down to Columbus and sadly told me that he had to give up his pet scheme because of the opposition of the vice-president.

While the commissioner was asleep after a night's trip, I wired the President of the Chicago, Milwaukee, St. Paul, and Pacific Railroad for a short interview for the next day. When the commissioner awoke, I had the president's answer naming the hour for the appointment. It is of common knowledge that one can approach someone in authority much easier than one can approach a small potato.

The railroad president did not know me, nor did he have the slightest inkling of the purpose of the interview. But, in ten minutes' time, I had sold the president on colonizing Hungarians in the Dakotas. The commissioner was then called in and authorized to proceed, first by hiring the man whom I recommended as their Hungarian agent, the young ex-superintendent of the mine in Kentucky. The president asked me what he could do for me in appreciation for my idea, and I told him that the only thing that I would like to have was not within his power to grant. What I wanted was a free pass on the railroad. However, granting of free passes to private persons is prohibited by the Interstate Commerce Commission.

A few minutes later, the president smiled and said that he could go one better and secure passes on every railroad under one condition. The condition was that I would permit the president to put me on the railroad's payroll as a colonization agent with a minimum salary of $100.00 per month. But he was quick to say that no work of any sort would be expected on my part; the action was just to show his appreciation.

I said that I could not accept pay for nothing, and he came back that I could continue advising the railroad and distributing their Hungarian ads. Further, if I found time to occasionally travel and observe farming in the various sections of the country, perhaps I could advise the railroad as to what nationality groups were best adapted to certain lines of farming and to certain climatic conditions prevailing along the railroad line from Chicago to Seattle. The president also assured me that, of course, if I found time to make such trips, my expenses would be cheerfully paid. Naturally, I jumped at the president's offer.

I permitted my name to be put on the payroll, and my income almost doubled. I decided to render all services I was capable of in return. Instead of traveling occasionally, I spent most of my time on the road, sending in my editorials from wherever I was and returning to Columbus not more often than once in ten days or so. I used this God-given chance to get really acquainted with our country by traveling through same in width and length. With the exception of some of the New England states, I believe that I know America as few native citizens do.

I became interested in farming and in every problem of the farmer. I familiarized myself with every branch of the industry, and I very much regret that farming is indeed an industry, instead of a style of life, as it should be. I learned the different aspects of the sugar beet industry verses the sugar cane industry. I learned how much more efficient the peanut grower of Florida is than the peanut grower in Virginia. I became familiar with the griefs and problems of the poultry men, the fruit farmer, the vegetable specialist, and the wheat manufacturer. If I may say so, I am one of the best theoretical farmers of our country. I visited thousands of farms of all kinds, in all parts of the country, and listened with interest to thousands of farmers during the three years of my vagabonding. As a result of the commissioner's kindness, I became especially familiar with every problem of the Northwestern Empire.

Once I was invited to join a traveling agricultural conference that started out from Chicago. Our special train carried twenty agricultural experts and at every stop, we were joined by other experts from agricultural colleges and universities and by practical farmers. Our train pulled into a station at 8:00 A.M. each morning, and each of us was taken around in the vicinity by a local farmer. Lunch was served at noon, and local farmers, businessmen, bankers, etc., started to verbally unload at the table. Farming, as well as industrial, business, and banking problems of the respective sections were discussed thoroughly.

Our train pulled out of each station about 2:00 P.M., only to stop within an hour at another station to repeat the performances. We lived on the train for thirty-four days, and we stopped in sixty-four localities. I received a priceless education for which I ever will be thankful to the Chicago, Milwaukee, St. Paul, and Pacific Railroad.

I was amazed to learn about the dependence of some sections of our country on the services extended by our common carriers. Europeans were somewhat suspicious when I pointed out the difference between American private enterprise going after business and building up business for themselves on the railroad in contrast to the state enterprises of Europe. As I traveled, I never failed to visit local libraries, museums, and book stores to satisfy my hunger about the early history of America. In fact, I wrote the history of our country for the *Magyar Bányászlap* in the form of interesting stories and adventures.

Meantime, I induced other individuals in other parts of the country to colonize Hungarians, despite the warnings of my colleagues that I was committing hari-kari by chasing my *Magyar Bányászlap* subscribers out of the coal mines and thus possibly causing the collapse of my newspaper. My colleagues were mistaken. A total of 70 percent of the Hungarians left the coal mines, and 70 percent of the subscribers of my newspaper, now owned and published by my nephew, Andrew Fay Fisher, had become farmers or industrial laborers. While the number of subscribers was declining fearfully, the decline was caused by the subscribers' moving to a better world fast, not by their being chased away from the coal mines. Since immigration is restricted, the days of all foreign publications are numbered. Hungarians enjoy the smallest immigration quota of all, some eight hundred immigrants a year. Therefore, Hungarian newspapers will die first.

When I called on so many farmers, publishers, and newspapermen, I made it my business to inquire about their political beliefs, their hopes, and their disappointments. I was interested in people.

I learned the impracticality of establishing a third political party, even in those days of William Borah[1] and George Norris[2] when the group of the liberal northwestern "sons of the wild jackasses" held the balance of power in Congress. I was not afraid the "Mr. Assistant President," Henry Wallace,[3] would succeed in destroying America's two-party system and follow the sad fate of European nations where the governments are always paralyzed by the necessity of constant coalitions.

I was happy to note how much more intelligent America's farmers were than the European peasants, albeit certain sections of the Old South were behind even the peasants. I came to the conclusion that, by and large, America had the most intelligent and best-informed farmers in the Northwest. Perhaps for this reason, the Northwest is more liberal than any other part of our country and has little traces of the nationalistic and religious prejudices of the South, East, and Midwest. Too bad that wars are always the best sources of intolerance because the Northwest also became somewhat infected with prejudice following the Second World War. Twenty years ago, some of the old agricultural pioneers were still alive, and the majority of them were of Dutch, Bohemian,[4] German, or other immigrant nationality. The unfailing pioneer spirit kept the Northwest clean of prejudice.

The Depression of 1930 forced the Chicago, Milwaukee, St. Paul, and Pacific Railroad to economize, their agricultural department shrank, and my happy days of travel were over. I stayed three years in Columbus, then moved to Cleveland where my close friends lived and where I spent most of my weekends. When I was ready to unload my printing equipment from the three trucks that I took to Cleveland, I found fifteen youngsters on the spot to help me. The youngsters were boys who were raised in Himlerville and whose parents had moved to Cleveland before me.

I was forced to stay home once more, and in order to provide occupation for myself, my nephew, and a very loyal friend who replaced the ex-postmaster in our organization, Mr. Nicholas Nadassy, I started to carry out one of my pet dreams. I published a new weekly newspaper in a form similar to *Nation*[5] and *New Republic*.[6] I devoted my newspaper to the interests of the offspring of the immigrants, and I named my newspaper the *New American*.

I was very sad to note, and one with open eyes could not very well help but to note, that in some respects our nation is willing to follow the Old

Testament by punishing the crime of foreign birth to the second and even third generation. There is no gainsaying the fact that Americans of foreign parents with foreign names very often, more than not, suffer almost as much discrimination as their parents did. The offspring are considered—to say the least—to have been born very much on the wrong side of the railroad tracks. They are subjected to the ever-present handicap of their births in their professions, businesses, and especially in their social contacts. I consider it almost beneath my dignity to point out that an immigrant citizen can measure the blessing of all that is American, whereas native-born Americans can never have as deep a love and as much reverence for America as the immigrants have. The children of the immigrants are also reared in this spirit of reverence for America. The *New American*, my spokesperson for the millions of innocent immigrant sufferers and my first English publication, was received with joy. It especially created much interest amongst the social workers. But, I timed the venture wrong. Immigrants and their families were the worst-hit victims of the Depression. Many of them actually went hungry during the last few months of the Hoover Regime, and they could not afford to pay $2.00 for the *New American*.

I went completely broke after a few months of struggle, and I became all the more broke because miners also were unable to pay for their *Magyar Bányászlap* subscriptions. Advertisers dropped out, and I was afraid that I couldn't keep the Hungarian newspapers alive. As a result of my last dollar's disappearance, I had to dispose of the services of both my nephew and Mr. Nadassy. My nephew returned to Hungary and took over a small daily newspaper. However, my nephew was unable to stand life in Hungary, and he returned to America a few years later.

I found that I now had to write the whole contents of three Hungarian newspapers, keep the books, and help some in the printing shop. I had to restrict my personal income to $10.00 per week, which was not much, even in those bitter months of the Depression. Luckily, I managed to pay my printers the usual $50.00 per week, and I never had to ask them to accept less. My men had stuck with me from Himlerville through Columbus, and their loyalty had to be appreciated.

Mr. Nadassy refused to be fired. He returned to me within a few weeks, helping as much as he could, and accepting $5.00–$6.00 or less per week. He kept the books, was a close friend of mine, and he knew best that even the $5.00–$6.00 meant a heavy load. To help more effectively, Mr. Nadassy learned to operate the Linotype.[7] Mr. Nadassy had been a noted Hungarian actor in his younger days. Later, of course, Mr. Nadassy always was as well paid as I was. Such was the loyalty of this man to me that when he died, I was notified when I was overseas that he left all that he had to me. The inheritance amounted to about $300.00 after burial expenses, and I spent twice that amount trying to trace his relatives in Hungary. Unfortunately, his relatives had been exterminated with the other hundreds of thousands of Nazi victims.

The hard times forced two Hungarian weeklies, one in Youngstown, Ohio, and one in Detroit, Michigan, to economize, and they made arrangements to have me print their issues. I furnished about three fourths of these newspapers' composition with news that I used for the *Magyar Bányászlap*. Only about one fourth of these newspapers was devoted to local items. This arrangement worked out very well, and I gained quite a large new public for my editorials.

It was not because I had to restrict myself to live on $10.00 weekly, but rather because I believed in Mr. Roosevelt's New Deal,[8] that I joyfully answered Congressman Sabath's[9] call to organize the Hungarian-born citizens for the New Deal Campaign. Perhaps because of my two inheritances, the ghetto and the poverty of Mátraverebély, and perhaps because, as an immigrant, I belonged to one of the under-privileged classes, I always had and still have the deepest sympathy for all underdogs of the world. The New Deal promised a new day for our own underdog. I jumped into the campaign with both feet and without accepting either pay or political jobs after the victory. I never regretted my activities in the campaign. In fact, I did the same in the Second New Deal[10] campaign, though I do not call myself a Democrat. I never belonged to either political party, and my newspapers were always independent.

Early next spring, I was on the move once more. I was induced by my Detroit customer to move to Detroit, and after loading up my equipment, I departed from Cleveland at the end of February, 1932. I transferred my bank account of $800.00 to a Detroit bank and, just when I was ready to pay the mover, the electricians, the bricklayers, the cabinet maker, etc., for the necessary alterations on my new Detroit shop, the banks were closed by Mr. Franklin D. Roosevelt before I drew a penny. I was standing there, a stranger in a strange city, dead broke, with bills and workers to pay. My

Detroit customer, the local publisher, was unable to pay his own bill, and I had to turn once more to my insurance policy for money. Some of my subscribers, and there were quite a few ex-miners in Detroit, came to my rescue by paying their subscriptions for years in advance from the few dollars they had earned on the Civil Works Administration employment. It is good to have friends when one is in need.

My newspaper business improved as several other Hungarian publishers of weekly newspapers discovered the advantages of filling three fourths of their newspapers with good syndicated material, and soon I was printing newspapers for the cities of Detroit, Flint, Youngstown, Dayton, Buffalo, South Bethlehem, Pittsburgh, and Trenton, in addition to my own three weeklies. I'd now reached a public much larger than any other editor or newspaper. Hungarian America had to contend with me once more! And I used my influence for what I believe to have been good purpose.

The Depression years tended to increase the victims of the Bolsheviks. Poor, ruined workers, divested of their hard-earned homes and other possessions, were easy prey in the hands of agitators. I was scared to note the increased number of Hungarians in Stalin's[11] fold.

I devoted most of my energy to fight Communism for various reasons, and the reasons were very selfish reasons. The first and main reason, of course, was my devotion for America and for everything America stood for, or should stand for, and my gratitude for the privilege of citizenship. I held, and still hold, that immigrants who become citizens of America have no right to have second thoughts by attempting radical changes in our Constitution.

I realized quite well that the very discriminations and abuses to which the immigrants are subjected are the best material in the hands of the demagogues preaching hatred not of the intolerant elements, but of America. Hatred begets hatred, and the agitations of the Ku Klux Klan and the ravings of Gerald L.K. Smith[12] and ilk are best recruits. Therefore, my second selfish reason to fight Communism was to save the Hungarian-born Americans from the consequences of attaching themselves to the Red groups.

I knew that employers would sooner or later try to weed out Communists from their organizations or, at best, would avoid hiring Communists when in need of men. I had no wish to see the more or less ignorant people, who knew little about what Communism really means, suffer for the misstep of being a Communist. I carried on a relentless fight against the Communists with the consequence that the Communist Hungarian daily newspaper at Cleveland had to fold up and publicly credit me for their ruin in their last issue.

But, I also suffered. All of the Communists' hatred and all of their energies were concentrated on the fight against me. Both the *Martin Himler Weekly*, and especially the *Cleveland Weekly*, became losing propositions.

I always have enjoyed a fight, and I did not mind to suffer the consequences. I did not mind, in earlier days, to lose much of the revenue of the *Magyar Bányászlap* when coal operators in West Virginia and Pennsylvania induced my advertisers to withdraw their ads from my newspaper when I was fighting the coal operators. Nor did I mind the filthy campaign of the Communists who used all the well-known tactics usually employed by them against their opponents.

I enlarged my Detroit printing shop, engaged in commercial printing as a side-line, and made enough money to stand the Communist fight and the losses. In fact, the *Magyar Bányászlap* itself failed to break even some years for I decided to quit going around to solicit advertisements and to lie about the usually small circulation of foreign weeklies. But it never occurred to me to kill the *Magyar Bányászlap* and to desert my coal miners. I kept faith with my miners until 1943 when I was called to America's military service. My nephew took the *Magyar Bányászlap* at that time and published the newspaper for the same sentimental reasons but made his living from other sources.

The National Recovery Act[13] of the early New Deal days enabled John L. Lewis[14] to re-organize the shattered union of the coal miners and, for the first time, to extend the same to the southern coal fields. In fact, the National Recovery Act handed Lewis the miners on a silver plate. And Lewis gobbled them up—lock, stock, and barrel. Lewis killed all vestige of democracy in the mining organization as he re-created the once-powerful union. Instead of members' electing district officers, Lewis appointed the officers with the excuse that the members were not yet qualified to elect officers. All the well-paid officers were depending on Lewis for their fat jobs, and the United Mine Workers became, for all practical purposes, Lewis' own personal property.

I knew Lewis and his methods well. I knew of his conventions packed with delegates from non-existing locals and his brutal methods of dealing with oppositions in previous years. I supported the several revolts of the

Twelfth Illinois District of the United Mine Workers and the anthracite miners with publicity and with cash. I knew the large number of the victims killed in these fights by Lewis' gunmen.

 I saw the large number of so-called organizers who, in reality, were hired guns invading the coal fields like locusts and intimidating the miners to have Lewis' men elected to local offices. I saw that Lewis forced locals to remit all monies to him with the locals' shares to be returned if Lewis were pleased with the management and officers of the thousands of local branches. I decided to do something about Lewis and his wicked ways.

CHAPTER 18

The American Miner

★

I PUBLISHED ANOTHER NEWSPAPER, the *American Miner*, a sister newspaper to the *Magyar Bányászlap*, and the *American Miner* was greeted with joy by many of the miners. Children of the Hungarian miners became my salespeople, and they sold close to ten thousand copies of the third issue.

Some of the lesser mining leaders, especially those not appointed to district offices to which they would have been entitled, took advantage of having an independent organ. A few even offered financial support which was refused by me. I also refused the financial support of one of my best friends, an old coal operator of Virginia who did not oppose unionization of his miners but who did oppose the dictation of John L. Lewis.

The manager of the largest Virginia coal mine became my friend in a strange way. When the *Magyar Bányászlap* was in its first year of publication, I attacked conditions then prevailing in Dante, Virginia, and I called the attention of the Virginia mine inspector to the lack of proper ventilation in the Dante, Virginia mine. I also criticized the Dante manager for lack of good drinking water and decent dwelling houses on the mining camp.

A few weeks later, the Dante general manager discovered me in the same train coach as he. The general manager introduced himself and told me that he had to spend $30,000.00 in remedying the situation in Dante. I was somewhat scared, for in those days of Baldwin-Felts mine guards, it was comparatively easy to deal with one who offended the coal operators. But my fear was soon dispersed. The Dante general manager told me that he respected me for fighting for my people and instead of harboring hard feelings, he would appreciate my calling his attention to future grievances in milder tones. We thus became good friends, and he often used the *Magyar Bányászlap* to secure more men.

This general manager offered to finance my newspaper to carry on the fight for a decently-managed union, but I never accepted money from any mining administration source, with the exception of the check from the Holden superintendent for personal services rendered. Realizing my own shortcomings, I engaged an excellent newspaperman to edit the *American Miner* and within a few short weeks, it was a very promising undertaking.

My Hungarian kids secured advertisements from local merchants with great success, and it seemed as though the *American Miner* would be a more powerful organ than the *Magyar Bányászlap* could ever hope to be.

But, it seemed to be my fate that the best-laid schemes and the best-built undertakings collapsed in my hands. The union goons went from camp to camp, intimidated the *American Miner* advertisers to withdraw their ads, and threatened my agents with expulsion of their fathers from the union or, in other words, with divesting the fathers of their livelihood. The coal industry became a closed shop with the National Recovery Act. I was broke once more within a year, and the *American Miner* was killed.

How could I fight John L. Lewis when the American government, as well as Congress, were impotent in dealing with him? More than ever, I was convinced that his existence was a menace to American institutions and that he maintained a dictatorship far more ruthless than Stalin's. While Stalin is subjected to some extent to the dictates of the Russian Politburo, John L. Lewis appointed the members of his Politburo. And while Stalin failed to establish his family as a ruling house, John L. Lewis forced his infamous Fiftieth District to accept his daughter, Catherine Lewis, as dictator-owner of the district with full authority over her subjects and the treasury of her district. Few Americans, in those interesting days of the Taft-Hartley[1] controversy, were aware of the existence of Catherine Lewis' domain. Her domain was part of the United Mine Workers, the organization of the coal miners, but the members of the Fiftieth District were intimidated into joining the subjects from various other industries not having the remotest connection with mining. One example of this was a large group of dairy farmers in Michigan who were black-jacked into joining Catherine's other subjects by her personal goons. I was not surprised to once receive a letter from one of my intelligent miner friends. The letter stated that "the coal miners were better off by having been terrorized by intelligent, if ruthless, coal operators, than by being terrorized by equally ruthless and unintelligent goons."

There were days during the hectic period of the sit-down mining strikes when I myself became discouraged. I had fought for mining unionization for twenty-five years, and I suffered financial losses for my beliefs. I finally had to ask myself whether our country was truly better off with such unions as the Lewis United Mine Workers.

Remembering the muck-raking[2] days of Teddy Roosevelt's attack on Wall Street, and noting the valiant fight of another Roosevelt against the same interests, I wondered whether America destroyed one privileged class only to create another one, the more dangerous one — the Lewises, Petrillos,[3] and Bioffs[4] — in its place. And who could deny that unioneers considered themselves a privileged class and terrorized government, Congress, the public, and their own members? Union bosses enjoyed luxuries far more luxurious than the President of the United States enjoyed, luxuries vastly beyond the dreams of the union workers. Educated young men slowly replaced the most objectionable union bosses, and, thankfully, the young men completely took over union management within a very few years.

When Mr. Franklin D. Roosevelt decided to break the tradition and seek the third term, I did not have it in my heart to campaign for him. I was beset by doubts about the wisdom of breaking with the past. I was raised in Europe, and I was fearful of a long tenure in office for the President. I was all for him, supported his candidacy in my newspapers, and voted for him, but I could not work for the dangerous third term.

When I was fifty years old, I thought that it was time to think of the future and old age. I had visited California several times, and on one trip, I took with me my faithful friend, Johnny Bako. Johnny and I decided that we eventually would locate in California. I called on Johnny, who was working as a gardener in Cleveland, and told him that the time to start for California was here.

We went down to Los Angeles and in the suburb town of Northridge, we purchased a ten-acre walnut grove for our prospective home. We decided to try poultry farming, and I left Johnny to build himself a temporary dwelling house, brooding house, and a chicken house to accommodate about two thousand laying hens. Both of us were to receive a nominal sum of pension in our old age, and we thought that if we could have a home and a small income, then we two old men could live in peace. The, for us, large tract of land was purchased because several of my friends in Los Angeles, Miss Fulop among them, expressed the wish to take over an acre or so and join Johnny and me in a small colony of five or six friends. Johnny and I were to retain only about four acres for ourselves.

I planned to join Johnny within about three years, after which the land would be paid for and a comfortable home built. I was looking forward happily to a contented life in a quiet place, far enough from the noise of the city yet amongst friends. I kept on sending Johnny as much money each month as I could. Johnny kept on building and developing the place, and he did wonders by erecting building after building, all by himself, in order to save every cent and to reach our goal as quickly as possible. Our poultry farm proved to be another rainbow, another dream I never attained. My mother died in 1925, and I had little incentive to return to Hungary. But, in 1936, after an absence from Hungary of over thirteen years, I wanted to see my sisters and brothers once more. All of us were getting old. I also wanted to see what was going on in Europe where Hitler was about to gobble up Czechoslovakia.

I had planned to spend eight weeks in Hungary, but within five weeks, I had had enough. I left Hungary greatly disturbed. Hungary was covered by Hitler's shadow even then. The first Jewish law restricting Jewish participation in business and professions had been passed, and the Germans applied economic pressure on Hungary to enforce the law much further. The Hungarian government introduced Fascist rules, and when I took to task the Hungarian Minister of Finance, he explained to me that Hungary was dealing mostly with Germany and Italy and, therefore, was forced to use the methods of Germany and Italy.

I was on the sea returning home from Munich as was arranged. I came home with the conviction that war in Europe would be unavoidable and that America would be involved. I devoted much of my time, upon my return, to organizing the American Hungarians into one body, something attempted several times before without success. The Hungarian priest of Detroit and I organized a local committee to take care of hundreds of visitors and then issued the call to Hungarians everywhere to attend a meeting. The meeting was openly opposed by the radical element and press and silently by the Hungarian daily newspapers. But, the meeting was held, and the American Hungarian Federation[5] was organized. The purpose of the Federation was to oppose the Nazi propaganda invading Hungarian America from the Old Country and to create a connecting link between our American government and the American Hungarians. The Federation was not a huge success but, as the secretary, I started to organize. As I tried to popularize the movement, I began to extend the services I always had extended to my coal miners to every Hungarian in America.

Two years later, in 1938, the so-called World Federation of Hungarians[6] of Budapest issued a call to Hungarians living in foreign countries to attend a meeting in Hungary. Hundreds of American Hungarians decided to participate with the belief that the World Federation was an innocent cultural association. I also was under the same impression. Three of the largest Hungarian-American fraternal organizations were represented by their respective presidents and priests at the meeting, and ministers attended to represent their parishes. I decided to look into this meeting matter myself. The meeting was a few weeks after the Anschluss.[7] I never will forget the joy in Vienna and the pride that the Viennese people took in the fact that theirs was the second city of the Reich.[8] As a result of the Anschluss, Germany now had a common border with Hungary, and the signs of speedy Nazification were demonstrated at every step.

Hungarian Prime Minister Béla Imrédy[9] was fighting hard for the second so-called Jewish law to force hundreds of thousands of Jews out of their professions and businesses; in fact, to divest the Jews of all possibilities of making a living of any kind. This same Imrédy was later proved to be of partly Jewish origin and was forced to resign by German authorities. Still later, the Germans furnished Imrédy with false documents, and during the German occupation of Hungary, Imrédy became a minister once more.

I visited Hungary at the time of the World Federation meeting and returned the visit paid me a few months before in Detroit by a high official of the Prime Ministry. I was received with open arms. The Hungarian official assured me that the new Hungarian government was ready and eager to financially support (subsidize) The American Hungarian Federation Press, and since I reached the largest number of readers through my thirteen weekly newspapers, I would receive a good share of this patronage. Oh, yes, this new Hungarian government also was willing to use men of Jewish origin if the men were willing to serve the Federation's purposes.

The official from the Prime Ministry was very much surprised, virtually dumfounded, when I told him in no uncertain terms that we former

Hungarians in America are now Americans, all of us, and we would not accept money from a foreign government. Further, I told him that if the two Nazi Hungarian weeklies—for to the everlasting shame of Hungarians, we did have two such newspapers—accepted subsidies, I certainly would expose them.

Noting the Nazi trend of the World Federation, I refused to participate in their meeting that was attended by the highest dignitaries of the state and that sadly was amusing due to American-Hungarian leaders fighting for the doubtful honor of being elected vice-presidents. One well-known Hungarian poet, a citizen of the United States for over forty years, objected in a great speech against his being left out of the vice-presidential choices.

A few days later, visiting American leaders were invited to a lunch given by István Antal, Chief of the Propaganda Bureau. I was seated next to Antal's deputy who immediately broached the subject of the newspaper subsidy and begged me to reconsider my stand.

Following lunch, I told Antal that I considered the World Federation's scheme an insult to every American, and we had a grand scrap. I later had the pleasure of arresting Antal in Austria in 1945 and sending him home to be hanged. One of the Propaganda Bureau lunch speakers, a Protestant minister, had enough nerve to tell the Hungarian dignitaries present that he was ashamed to show Hungary to the youngsters he induced to accompany him to the land of their fathers. The minister gave the dignitaries a piece of his mind about the treatment of the Hungarian peasants. But, by that time, most of the American Hungarians had been dined and wined by ministers, counts, or what have you, and by aristocrats who would not spit on these American Hungarians before they emigrated to America. The foolish Hungarian Americans turned on the brave minister. The lunch host responded to the minister by saying that Hungary, as well as Germany, had to forego butter by choosing cannons because they were preparing for war. The lunch guests became carried away by their enthusiasm for, in the Nazi atmosphere, the guests themselves had become more or less "Nazi-fied."

The presidents of the Hungarian-American fraternal organizations acted on secret instructions received in Hungary and, upon their return to America, the presidents joined the American Hungarian Federation. The presidents forced me to hand them over the majority on the American Hungarian Federation Board of Directors by pointing out that the Federation was weak and that the support of their organizations would make its existence permanently secure and able to maintain headquarters in Washington, D.C. I knew that the presidents' antagonism would kill the Federation, but I had no choice except to hand over the majority. I was replaced as Secretary of the Federation within a few weeks.

The president of the largest fraternal Hungarian organization was made Chairman of the American Hungarian Federation Board, and a few words about him and his dealing with immigrant people will not be misplaced in this book. The president, an old resident and citizen of America, posed for years as a radically liberal well-to-do merchant who was a teacher in the Old Country. Several years before, I had selected him as best fit to be president of the American Hungarian fraternal organization which was composed mostly of uneducated Hungarian workers. It was common knowledge that I built up the president in my newspapers and caused his election as president of the fraternal organization. He was very grateful to me for years and through him, I exerted quite an influence on the Hungarian public life. The influence of the priests' and ministers' visits to "Nazi-fied" Hungary compelled one of the priests to write a letter to the new chairman. The letter objected to my replacement as Secretary of the Federation by a man unknown to American Hungarians and unfit for the job. The new chairman's answer was as follows: (I have the letter in my hands.)

> I know that Himler is the most capable man for
> the job, and that Professor X does not know
> anything about American Hungarians, but, at least,
> the professor is a Magyar, not a Jew.

The Nazi poison was spreading all over Hungarian America. Two of our weeklies were openly Nazi, and several others were slyly infecting their readers. Were it not for my newspapers' growing anti-Nazi participation, the entire half-million Hungarians of America would have been infected. Willing followers of the Nazi propaganda were a very few Catholic priests and more than a few Protestant ministers with their eyes over on the Old Country, and I had to fight both subversive elements in my newspapers.

When I was kicked out by my friends as Secretary of the American Hungarian Federation, I remained on the board of directors to check their doings. And their doings were worth checking.

Prior to America's entering World War II, I proposed that the Federation have one of the Allies drop a million leaflets to the Hungarian people and call on them to sabotage in the name of their relatives in the United

States. My successor as Federation secretary, a professor at a Catholic university, was aghast at my suggestion. He said to me, "Mr. Himler, suppose a few Hungarians take the leaflets seriously and, as a consequence, they will be hanged!" This response was given to me when tens of thousands of Hungarians were cut to pieces on the Russian front. Another professor, one of the pillars of the Federation, wrote an article endorsed by the new secretary. The article stated that the Federation was neither pro- nor anti-German. This statement was made on the verge of America's entering World War II.

Once, a special meeting of the Federation board was called, and a man from the Hungarian State Department was the meeting speaker. The man "advised" the Federation to endorse Tibor Eckhardt, President of the Smallholder Party[10] of Hungary. Eckhardt was brought to America by the British, and his aim was to organize a resistance and independence movement by Hungarians living abroad.

Tibor Eckhardt and his party were loyal opposition of the Horthy Regime and had as liberal a programme as was permitted in Hungary. But Eckhardt himself, in his younger days, was one of the leaders of the "Awakening Hungarians," the white terrorists who organized themselves after the collapse of Béla Kun's[11] Communistic Regime in 1919. Liberal American Hungarians were not elated by Eckhardt's visit. When I questioned Eckhardt as to whether his support was the wish of the State Department, his answer was that supporting him was not an official wish, but American Hungarians could do a lot worse than endorsing him.

Eckhardt was endorsed, but his movement was subjected to such a vehement and concentrated attack by the Czechoslovakian propaganda bureau and radical Hungarian American groups that it died aborning. Finally, the State Department was forced to issue a denial of the implied or direct support of Eckhardt.

I was mad because my newspapers had stated publicly that Eckhardt was introduced to the Federation by the State Department and with its blessings, and the ensuing denial put the Federation into a very peculiar light. When I called on the Hungarian State Department to object to this machination, I was told that bearing the false accusations must be our newspapers' contribution to the war efforts.

Mr. Eckhardt retired from the limelight and lived quietly in New York. But, I was told through the Federation's secretary that Eckhardt exerted more influence on the American Hungarian Federation than all of its members put together.

After this chicanery, and with the increasing Nazi tendency of some of our Hungarian leaders, I became disgusted, resigned from the Federation Board, and devoted my time to my newspapers. I hoped to join Johnny in California about 1942. Then came Pearl Harbor. And, of all the criminal fools of the world, Horthy's government declared war on the United States.

I wired to the predecessor of the OSS and offered my service to my America. I did not ask for office space and a chair in Washington but advised the powers to send me as a Private to Draza Mihailovic,[12] (Tito[13] was not in the picture then.), from where I expected to slip over to Hungary and organize resistance movement in the country. I knew hundreds of Hungarians who used to live in the United States, and I was confident that I could do something worthwhile with them.

The answer from the OSS came, not in the usual mimeographed form, but in a personal letter advising me that I would be called if and when my services could be used to the best advantage. And the call did come — almost two years later!

It was two very tough years.

Every Armistice Day made me sick, for I felt that I also should have been in America's service in the First World War. My task in the two years after Pearl Harbor was almost unendurable. I had to write editorials urging the poor old Hungarian women to send their four, five, or six sons to the war smiling, and I felt like two cents doing that in my well-heated and safe office in Detroit. The children of my friends, my children, went one by one, and my face was burning whenever I passed a youngster in uniform on Detroit's streets. Three boys were called from my shop, and I still was sitting there.

I wrote to Washington repeatedly without results until, finally, in September, 1943, I was called over the phone and was told to arrange my affairs because I soon would be called to serve America. Less than twenty-four hours later, I sold my shop to the first comer at about thirty cents on the dollar, and I killed my *Cleveland Weekly*, and the *Martin Himler Weekly*. I gave the *Magyar Bányászlap* to my nephew, its present editor and publisher, with my blessings.

I went to California to say good-bye to Johnny and to assure him that all of our plans would be carried out on my return. I made arrangements to have a small house built on our ranch for my other loyal friend, Mr.

Nadassy, who was ill and unwilling to serve as the new owner of my printing shop.

I found Johnny in difficulties. When he finished the buildings, he started to raise chickens, and he purchased two thousand very expensive baby chicks. They drove him crazy. The little cannibals were eating each other and made Johnny nervous, so much so that I advised him to drop the business and attend only to the orchard until my return. But he would not hear about it. He wanted to show profit in my absence.

The wire from Washington arrived on December 22, and I was on the train the following day with much misgivings. I knew that I would have to pass a physical examination. I was over fifty-five years old and healthy, but suffered every now and then with sinus trouble and other small irritations of my age. I did not know what my assignment would be, and I was worrying that perhaps the task would be beyond my capacity. I kept worrying about getting along with other people, especially with much younger people, after having spent decades of being a sort of a lone wolf and an introvert by nature.

The day after Christmas, I was in Washington, D.C. I spent Christmas on the train. Four days later, I was sent to a training school. I mentioned earlier the thrill of a lifetime with my first sight of the Statue of Liberty in 1907, and I then said that by the grace of God I enjoyed a similar thrill thirty-seven years later. The second thrill came with the GI uniform that I was privileged to don before going to the first training school. I may be a sentimental old fool, but I had tears in my eyes. I felt that I had reached the height of my career because no greater honor can befall any man than the honor of wearing the uniform of a private of the United States.

I attended three GI training schools and was happy to experience the comradeship and the friendship revealed in all three of the schools. I did my best to fall in line with my new and young buddies, and I was overjoyed to see that I was accepted by them. For the first time in my life, I ceased to be an outsider.

We were taught many stunts in those training schools, and we had a grand time in the evenings. I wondered whether I were really an incurable introvert or if perhaps I were still laboring with the inferiority complex I had carried ever since I was a child in Mátraverebély. I enjoyed the company and the fun of the youngsters, and they were appreciative of the so-called wisdom of the few older men amongst them.

Six weeks later, I was sent back to the third school as a chief instructor. I was ashamed to look into the mirror when I looked over my students who were old newspapermen from the Far East, university professors, and writers, all of them in towering heights above me in knowledge and intelligence. I was unhappy about fooling my time away in a school when I was impatient to receive my assignment and go overseas.

The term in my school was one week and, at the end of the first week, my order came. Since I was seasick most of the time on ships, I was very glad to see that my order told me to fly to Cairo[14] and proceed from there to Istanbul. The day before I left, I was told what my duties would be. I was somewhat scared, as well as honored, by the responsibility and confidence placed in me.

When I sold my printing shop, I simply stated that I was retiring to my farm in California. Only four or five of my closest friends knew that I had entered the service and gone overseas. I did not seek publicity and glory, for I did only what thirteen million other Americans were glad to do.

I got acquainted with four of my colleagues in the same service while we traveled to Miami. All of us were unhappy and impatient to be held up in Miami for seven days. But, I made the best of it and spent a glorious week on the beach.

The seventh day in Miami, my new friends and buddies left. I was alone again, unwilling to mix with men in other services lest I breach the security prevailing in OSS. I felt like crying. The same night, I received a place on a small plane carrying airplane engines to India. I was then on my way to Natal, South America, with a stop-over in Jamaica. In Natal, I caught up with my friends, and we were kept in Natal for eight long days. We felt every minute of every day that we were missing something and that the very war effort was losing by our delay.

We "took in" the small Brazilian town almost every day, and I have a fair idea of the life in a South American coastal town. I also was introduced to the ingenuity of our Quartermaster Department, and I was very proud to note that our boys were furnished with the best of everything, no matter where they were. Our GI's had well-planned barracks, a well-supplied PX, one-day laundry service, movies, and what have you. There were two movies on the American compound, one conducted by the Army, one by the Navy. If for nothing else since then, I believe in the unification of America's armed forces.

Eight days later, my friends and I left Natal. Since we were jammed into a plane like sardines, we were lying on each other for the overnight trip to the Gold Coast of Africa. One of my buddies, and still my best friend, can't forgive me to this day for waking him up several times by placing my feet (and shoes) on his stomach on this trip.

While watching a beautiful daybreak the following morning, my buddies and I sighted the little flyspeck of Ascension Island, and we made bets that our pilot would be unable to land on it. We were mistaken, of course, and we were served an excellent breakfast by soldiers. Shortly after noon, we landed at Accra on the Gold Coast of Africa, and we were kept there another eight days. We traveled with No. 3 Priority, and there were too many brass hats and diplomats with higher priorities who left Accra first. When my time to leave Accra came, I felt that I was well re-paid for everything that I might be called to do in the future. I was more grateful to America than ever before. I had had only a peep into the mysterious Dark Continent, but the impressions and information that I received would be enough for an average lifetime.

My buddies and I were not encouraged to talk with Accra natives, but what GI ever obeyed all direct or implied orders if he had a chance to break a rule or two? I went to the native town of thirty thousand every day, and I especially enjoyed the indescribably colorful Sunday when an additional twenty thousand natives came in from the nearby kraals. I made friends with traders selling curios to American and British soldiers, and I visited the native school at Accra, as well as the schools conducted by American Methodists. I talked with Greek and Hindu merchants, and I had quite a glimpse of the so called "white man's" burden that is loaded onto our British cousins. I came to the conclusion that carrying the same burden is not a bad and losing game at all.

The Accra natives were friendly to Americans who did not deem it beneath their dignities to talk to natives or even to treat them to a cigarette or two. Some of the natives talked quite freely, not always in complimentary terms, about their white masters. The natives were an intelligent group who were unhappy about the artificially high cost of education (meaning grade school education) and who bemoaned the fact that few of them were permitted to learn even a skilled trade. The junk sold to the natives at excessive prices would have been thrown out of an American ten-cent store fifty years ago. I learned, to my great surprise, that none of the native men would marry, or even touch a girl, before they were twenty-one years of age, which may be the explanation for the natives' fine, strong stock.

I could not help but compare the British's treatment of their colonies, or at least some of their colonies, with what I had read from time to time about America's rule over the Samoa Islands, about our American schools in the Philippine Islands, and about the constant deficit in America's governing Puerto Rico and the Virgin Islands. I thanked God that I am an American.

Eight days later, my friends and I were on our way to Cairo, a twenty-eight hour trip by plane with four stops for re-fueling. On this and on every succeeding trip, of which there were many in my close to four years of service, I always was pleasantly surprised to experience the care of the Air Transport Command for its passengers.

No matter what time of the day or night, hot food was waiting on the table for the Air Transport Command passengers upon arrival at any port. When stop-overs were for the night, good billets were provided immediately and were always the best available—the best hotel of a city or, on far-away compounds, large, convenient, and healthy barracks. I was billeted in a large barrack at Accra, and there was a ceaseless poker game going on when I arrived at the barrack. The game continued day and night and was still going without any let-up when I left the barrack. New arrivals took the places of the departing players, and planes were coming and going day and night. For all that I know, the poker game went on until the end of the war.

Passengers on our long trips also amused themselves by playing poker or reading. Meals were served at every five to six hours at stop-overs, but cold fruit juice, sandwiches, and fruit always were served on the planes of the Air Transport Command.

Cairo, the crossroad of three continents, was loaded to bursting with Allied troops and, at first glance, it seemed a cosmopolitan city devoid of the pure characteristics of the Orient. I was afraid that I would have to leave the fascinating city before having a chance to explore its byways, but within twenty-four hours, I knew different. There was a jurisdictional fight between the chiefs of the OSS of Cairo and Istanbul, and since I was under the jurisdiction of the former while in Cairo, I was told that I would remain there permanently. My logical station would have been Istanbul, which was still in touch with Hungary through her Allies of Romania and Bulgaria, and still full of diplomatic and secret envoys with some agents of the pitifully small Hungarian newspaper organizations of underground also present. But it was "just one of those things," and under the red tape rules of the Army, I had to set up my shop in Cairo.

My work was interesting, and I did the best that I could under the circumstances. I explored the city in my free time and fell in love with the meeting place of East and West, a place full of immense riches and unspeakable poverty rubbing shoulders at every step. I never ceased to wander and wonder at the bazaar, and, taking advantage of my civilian clothes, I spent most of my time in parts of the city "out of bounds" for Allied personnel, the parts of the city that were clear Arab sections.

I grew very fond of the patient, good-natured, ever-smiling Arabs. The friendliness of the poor city proletariat, as well as the fellahin, the Arab peasant, was a revelation to me. I saw more smiling faces amongst Arabs who never knew the luxury of a bed or even a roof over their heads, for that matter, than one can see on a nice spring day on New York's Fifth Avenue. As hard as I tried to discern whether their contentment was due to their climate or to their religion, I have not succeeded to this day. One could see the Arabs walking the streets, hand in hand, singing most of the time what sounded like weird Oriental tunes that remind one of the wind of the deserts. Or, the Arabs were seen working in their open shops as late as 10:00 P.M. – 11:00 P.M. with tools discarded by the West five centuries ago (British colonial policy?). Then, the Arabs go to the ever-present cafes for a few hours of friendly gossip. One can see the Arabs taking their siesta on the sidewalks on a piece of paper. While some Westerners have nothing but contempt for the Arabian poverty, one can't help but be touched by their child-like informality. The black cover on the Arab women (all women) is depressing, but once one can get a glimpse at the colorful dress under the black robe, one gets accustomed to visualizing the women at home when the black cover is discarded.

It can't be said that ignorance is the source of the Arabs' contentment for the slummy sections are alternated by the luxurious palaces of the wealthy, and beggars sleeping on the sidewalks are often disturbed in their sleep by the promenading wealthy in their be-jeweled silk robes. Dozens of people who never had a home slept nightly a few feet from the famous night club, said to be owned by the king and visited daily by people having pearls and jewels worth ten hundreds of thousands of pounds on their bodies. Thus, the filthy rich and the filthy poor rubbed shoulders as it were. I've always been irritated that American officers and enlisted men fall in with the British in calling the Arabs "wags," our well-known habit of looking down on "natives" wherever we go.

Once, when I was hitch-hiking on a Flying Fortress[15] from Palestine, I was lucky enough to be dumped off at the small airport of Deversoir on the banks of the Suez Canal. Instead of going to our small detachment for billet, I spent a delightful night in the home—or rather, in front of the home—of a fellahin. Spending the night in this way was a court martial offense. The home consisted of a mud house of about eight by eight feet with a tent of the same size attached. The mud house and tent were home to a family of parents and five children. The father and two of the children spoke some English, and we talked into the early morning hours. I received the pancake-like bread, goat's milk, and dates for supper, and the same for breakfast.

The family was comparatively rich and had about half an acre of land, two goats, and an occasional wartime employment with the British at the rate of 10 Piasters, ($.41) a day. The father told me about the hard times before the war and about small Piaster, ($.02) wages paid by the Egyptian landlords. I had to force half a Pound into his small child's hand for my host refused to accept payment. But, he took my GI shirt eagerly in exchange for the dates he insisted that I take with me for the trip to Cairo. And with the shirt, my host did become rich, indeed. The shirt was worth about 4 Pounds, the price of two good goats. I could not explain my shirtless appearance at the detachment next morning and instead of requesting transportation, I hitch-hiked with a British lorry[16] to Ismalie and took the train for Cairo.

While in Egypt, I had a good chance to compare American troops to those of our Allies, and I was proud and happy to note the absolute superiority of the American GI's. Since I was reared in Eastern Europe, I had the exaggerated idea about the superior intelligence and education of the British. I certainly was disappointed at the level of general intelligence and the manner of the British troops. Intending no offense, I do not hesitate to state that our American Negro boys from the deep South had better manners than the British troops and, the only Allied boys who came anywhere to our own were the Australians and New Zealanders, with the Palestine Brigade second best.

I derived immense joy from the behavior and the kindliness of the American boys during my entire stay in the service. When at home, we are inclined to judge our youngsters by the few rowdies at street corners or the few loud-mouthed students in schools and colleges. We sometimes lose sight of the vast majority who do not "advertise" but who, instead, study and work quietly. If there were any group of military individuals to criticize, the group was the officers. Authority and income were thrust on them without their having been prepared by previous life to exercise authority.

I collapsed in my office in May, 1944, and the physicians of our detachment ordered me to the wonderful army hospital at Huckstep, not Hospital King Farouk which was located on the desert about twenty miles from Cairo. The hospital at Huckstep was one of the best; it had very few parallels in the United States. I understood that it was manned and equipped by the Jefferson Hospital of Philadelphia.

While I was a patient at Huckstep, I learned an interesting bit about the business methods and general attitude of the Egyptian government. It was no secret that the Arab world, especially the Egyptians (not to mention the unspeakable Grand Mufti[17]), was wishing and hoping for a German victory. When Rommel[18] was about to reach Egypt, the British tanks about the Palace kept order only in Cairo. A large sign was put up welcoming Rommel on the main boulevard of Alexandria, and a small German flag was ready under the burnous[19] of every Arab. And, when I had learned a bit about the British rule of the Middle East, I couldn't very well blame the Arabs for their German sentiments.

The hospital equipment was unloaded at the time, and in the great hurry to evacuate the few American troops when Rommel arrived in Egypt, the equipment was left in Egypt until a short while later when we were able to return. The Egyptians then claimed the equipment as "salvage," and we had to buy it back from them.

While I am opposed to naming ex-members of the OSS, I have to mention Colonel Charles Vanda of Hollywood, California, with whom I was fortunate to work before I was stricken with the gall stones. Colonel Vanda took me to the hospital and did my work, alone, for six weeks. Colonel Vanda drove out several times to the hospital in the heat of 110-125 degrees to visit me, and he showed kindness that the experience of receiving made life worth living. I was unaware up to then that I had gallstones, a very painful asset, until my gallbladder was removed. My stay of six weeks in the hospital was a time when I, as well as the hundreds of other patients, received care and consideration that money can't buy in the United States.

When I left the hospital, I was very sincere when I told the chief surgeon that it was worth having my belly cut open to observe the kindness, patience, and consideration given to the sick and wounded by army physicians, nurses, and orderlies. Later, I had the same experience in the army hospital at Bari, Italy.[20]

While in Cairo, I gained some inkling about international politics in the Near and Middle East. I was amazed to see British troops surrounding the palace of the Greek king in Cairo to save him from attacks by the green troops[21] stationed in the city.

When I made my various business trips to Palestine, I had a chance to talk with both Jewish and Arab leaders and to note the wonderful accomplishments of the Jewish settlers. The Jewish settlers were responsible for the planting of crops in Palestine and for bringing sanitation into the midst of the filth of the region. No matter how hard I tried to judge the Palestine question without prejudice, I couldn't fail to indict the British for their past crime in creating the antagonism between the Jews and Arabs and for breaching their solemn promise to the long-suffering Jewish people.

I would have been very happy in my work if it were not for my constant worries about members of my family in Hungary. My job forced me to be very well informed about events in Hungary, and I received and read Hungarian newspapers that gave me a day-to-day picture of the unfortunate Hungary. I knew that after the German occupation on March 19, 1944, Hungarians were forced into ghettos and subjected to brutalities and humiliations beyond the conception of normal people.

I read the orders of the Nazi government to deport all Jews from Hungary and knew only too well what it meant. But I was hoping against hope that the Hungarian government would not be insane enough to disrupt war efforts or even to spare rolling stock[22] for the deportation of 800,000 people. I could not believe that the Hungarians whom I knew and with whom I was raised could have been brutalized to that extent. I kidded myself that the orders were merely issued to satisfy the Germans. Alas, I did not know how easily and to what depths a nation could sink when brutality is connected with plunder.

A few days after the collapse of Romania, I was ordered to Bari, Italy, and I knew when I left that if I lived long enough, I would see Egypt again as a tourist. It is the hardest part of the service to suddenly leave comrades and friends behind, in most cases never to see them again. But at this time,

I was glad of the change. Cairo was remote from the front, but war was still raging in Italy, and I was hopeful that my future activities would get me nearer to the front and to the actualities of the war. In this hope, I was disappointed.

America knew that the war was about to come to a climax and that Hungary—part of the "soft under-belly of Europe"—couldn't last much longer. Of course, we did everything we could to bring her surrender as soon as possible. Several Hungarians—leaders in the so-called "resistance groups" which did mighty little resisting—escaped from Hungary by plane, and some of our own men were in Hungary. Thus, we had a clear picture of the disintegration of Regent Horthy's Regime. I learned from the escapees and our own men the sad truth about Hungary's treatment and the all but complete extermination of Hungary's Jews. I spent many sleepless nights visualizing a sister or a brother of mine taken to the gas chambers.

When I first received my OSS assignment in Washington, I knew that I would be amongst the first Americans in post-war Hungary. I was happy to anticipate that my own little personal history would repeat itself and that I could again bring relief and help to my family and friends in their need as I had done in 1918-1919.

At 11 P.M. on October 15, 1944, I was aroused from my sleep by one of my colleagues who gave me the good news that Horthy surrendered. I got up at once to pack my duffel bags and hoped that within hours we would be on our way to Hungary.

There was a group of fifty of us organized for the purpose of help to Hungary, and we were trained and briefed to invade the city, set up shop, and fight Austria and Germany from their borders. Our group was alerted the next morning, and alerted we stayed for several months as we got a taste of the Iron Curtain of our own ally, Russia, who failed to grant the clearance urgently requested by Washington.

Months later, and with the final realization that Russia would not permit America to enter Hungary, my unit was ordered to Salzburg. Then, during the exciting days of the German collapse, our group established headquarters in that town. I was disappointed. My activities were always concentrated on Hungary, and I did not feel that I could render proper service in Austria or Germany. I requested my release. The war was over, and I felt justified in terminating my services.

I was requested to stay in Rome for a few weeks more and somewhat later, I was ordered to do a certain job in Milano.[23] While there, I was suddenly called to Headquarters at Caserta,[24] and I took it for granted that it was to receive my Air Transport Command for home. Instead, I was shown a cable from Salzburg stating that the United States was faced there by hundreds of thousands of Hungarians. I was ordered to proceed to Salzburg at once and to take charge of, or rather to re-organize, the Hungarian section.

I was happy with the prospect of being active once more, and I enjoyed the wonderful drive thru the Brenner Pass,[25] the most beautiful tour of my life. Americans touring Europe commit a crime against themselves if they fail to drive through this scenic pass.

I went to Italy with prejudices against the Italians. I came to America during the heydays of the Black Hand,[26] and I always steered clear of associating with Italians. I went to Italy while the county was still fighting America for the second time in twenty-five years. In Southern Italy, the people were filthy, dressed in rags. American troops were warned by large signs erected by the United States Army that some Italian villages were full of lice and full of thieves.

Six months later, I loved the Italians, and if I should ever have to live in exile in Europe, I certainly would choose Italy. Yes, Italians were thieves, for they were hungry, all of them. They were filthy and stole everything that was not nailed down. But they were human beings, and they had a sense of humor. They took their poverty and hunger with a smiling stiff upper lip. The dirty little urchins sold newspapers, shined shoes, or begged on the streets, bare-footed and hungry on cold winter mornings. But they sang arias from *Tosca*[27] as they begged. One can't attend operas in a small Italian town without falling in love with the cold and hungry beggars who often sang along with the actors and entertained the audience between the acts.

No one can see the hardworking Italian peasant husbanding his cruelly small parcels of land without respecting him. The Italians were Bohemians in the true sense of the word, and they had a heart. Jews, having gone through German, Yugoslavian, and Italian concentration camps in succession, have nothing but praise for the Italians. Italian police and carabinieri[28] in innumerable towns and villages warned Jews two or three days ahead that orders were out to round them up and advised the Jews to hide themselves.

When I arrived in Salzburg, it took me several days to "take in" the Hungarian problems. There were about 700,000 Hungarians on our hands, including remnants of the whole Hungarian army who now were our prisoners of war. There were over 250,000 liberated slave workers with a pitifully small group of Jews who were left alive only because the Germans were limited in their noble work during the last hectic days of the war by the capacity of the gas chambers. Hungarian industrial plants with all the personnel had been removed to "safe" German territory before the Russian invasion, and now the personnel of the industrial plants had come to Salzburg. The staff and students of universities and colleges, state and county administrations, the gendarmerie, and tens of thousands of individuals who were afraid of the Russians left Hungary individually and also came to Salzburg. It took various branches of the hastily set-up American military government weeks to classify everyone and to bring about a semblance of order in the tremendous anthill of humanity.

My offices were besieged day and night by desperate Hungarians trying to get information about members of their families. When slave workers were taken, and when the Nazi-Hungarian government ordered removal of manufacturing plants before the Russians arrived in Hungary, members of families were separated. The general chaos resulted in millions of people wondering and worrying whether they were widows or widowers or still had a mate somewhere. They wondered if they were orphans or still had parents, and they wondered if they were now parents without children.

Indeed, in that year of 1947, there were still tens of thousands of people in Europe seeking information about members of their families, for tens, if not hundreds of thousands were unable to send or receive information through the Iron Curtains of Russia, Poland, Yugoslavia, or the Russian-occupied zones of Austria and Germany. Even families, or man and wife running away together from Hungary during the last stages of the war, were often separated from each other in the mad rush and were unable to contact their loved ones for months, thus not knowing whether the loved ones were dead or alive. And many were the dead! Multiply America's problems by ten, for we had Italian, Austrian, Polish, Romanian, Czechoslovakian, Yugoslavian, etc., refugees, and you may have a hazy idea of what our military government was up to.

Finally, some sort of order was established amongst the ever-moving millions. Camps, shelter, and food were provided for the refugees, and I turned my attention to the members of the last Hungarian government. We wanted to have a clear picture of Hungary's relationship with the Germans. We desired to know the extent of her participation in the War of Aggression and in the colossal crimes against humanity. And, we wanted to discover the responsibilities of individual members of Hungary's last government. I was greatly honored one day while I was so occupied.

When I returned from a trip to Berchtesgaden[29] on a Sunday afternoon, I was told in the mess hall to hurry to my office to meet some distinguished visitors. Distinguished visitors did not mean much in my life during those days for we were used to showing the door several times a day to barons, counts, princes, generals, or what have you, who came to offer their services or to seek some business possibilities, some concessions, or just protection. Grudgingly, I drove to my office, and the CQ immediately announced Justice Robert Jackson,[30] General William J. Donovan,[31] and Colonel John Harlan Amen.[32]

I was embarrassed to realize that these three great Americans and really distinguished gentlemen had had to wait for me, and I was even more embarrassed when Justice Jackson told me that he came over from Nuremberg[33] to see me. After all, General Donovan could have ordered me with one word to Nuremberg, Cairo, or New York. And, when Justice Jackson told me the object of the visit, I received the first lesson of my life in what real Americanism means.

I had learned during my first few days in America while I was in Portsmouth, Ohio, that public servants in America are really the servants of the public, unlike in Hungary where the public servants were over the people who supported them. While I was a peddler, I learned that America stands for freedom of trade. When I was converted to Catholicism, I found no official red tape, and I realized that freedom of religion truly exists in America. As a publisher, I demonstrated America's freedom of the press to my own satisfaction. And in my conception of what Americanism means, I always have included all of these blessings.

Thirty-eight years after my arrival at these American shores, I learned another and most beautiful aspect of Americanism. The reason for the visit of Justice Jackson, General Donovan, and Colonel Amen was to tell me that my section would arrest about one to fifteen of the most notorious Hungarian statesmen, all of them criminals. I was to interrogate them and select the most prominent of them for trial by the International Court at Nuremberg.

I pointed out to Justice Jackson, General Donovan, and Colonel Amen that we could not even scratch the surface of the vast army of the arch criminals if we arrested no more than fifteen Hungarian statesmen, for we had hundreds, even thousands, of the criminals in our midst. Even then, I had the bulletin of the new Hungarian government which named over four hundred notorious gangsters, most of whom were hiding in American zones. I requested permission to arrest all of them.

I was startled at the reaction of Justice Jackson.

He told me briskly that I was supposed to be an American and as such I ought to know that from the beginning of the existence of our free country, the United States always granted asylum for political refugees.

I tried to point out that the men named as criminals by the Hungarian government were really criminals under America's conception as to what war crimes and crimes against humanity were, and I cited information from defending such crimes from the circular issued by our own American government.

The Justice still disagreed.

He held that handing over the political leaders of a past regime to the revolutionary courts of a new regime would violate America's tradition. General Donovan, Colonel Amen and I then listened to Justice Jackson deliver the most logical and beautiful dissertation I ever was privileged to hear about America and Americanism.

I never was as proud to be an American as then, but I knew that Justice Jackson was applying his logic and his heart at the wrong place. And with his permission, I started to read the names on my list, describing their crimes one by one. I described the crimes of mass murder, of enslavement of hundreds of thousands, of hideous plunders, and of violation of everything sacred under the code of civilization. I also reminded Justice Jackson that Radio Moscow was even then charging day after day that the United States was providing safe haven, nay, that we were welcoming into our midst hordes of Fascistic war criminals. I told Justice Jackson that his attitude would justify the Radio Moscow vile accusations.

General Donovan came to my rescue.

General Donovan had been listening quietly for half an hour to the debate between Justice Jackson and myself. However, when I mentioned the Russian propaganda, General Donovan cut in. He complimented me by telling Justice Jackson that, "Himler knows more about these criminals and about the Hungarian situation than we will ever learn or want to know. He is right about the Russian propaganda."

Colonel Amen also spoke up in favor of cleaning up the criminal element. I then was ordered to arrest and hold everyone named as war criminals by the Hungarian government, provided that I also found them guilty by America's own definitions. I was to arrest everyone whom I held guilty of war crimes, even if the guilty were not listed as criminals by their own government. I then was to select twelve or fifteen of the accused, the most prominent ones, for trial at Nuremberg.

Today, when liberal Americans are afraid that a witch hunt is about to be staged by the Federal Bureau of Investigation because some disloyal people are about to be discharged from government services, I feel quite confident that no injustice will be done. As long as Justices of highest court are as zealous of human rights as Justice Jackson was of the privileges of men who were our enemies a few days earlier, I have the utmost confidence in our federal courts.

CHAPTER 19

Playing God

★

I SIGNED OVER THREE HUNDRED arrest warrants on the same evening that Justice Jackson, General Donovan, and Colonel Amen left my office, and I am not ashamed to admit that it was a labor of love. As an American, and as a human being with some decency, I was horrified at the sight of Dachau.[1] As an individual, I was tortured by the fear that I might be stepping on the scattered bones of one of my own sisters or brothers, and as a man of Hungarian birth, I knew that the Nazi gangsters were mostly responsible for the infamy of the Hungarian nation. I thus had little sympathy for them.[2]

I was confronted with horrors beyond the imaginations of the average American throughout my activities in Salzburg. Many a sleepless night I doubted my own senses, doubted whether I saw and heard the awful things I did, and wondered if I slowly were going insane. I had nightmares even when awake. As I dealt with the criminals and their crimes, I often had to question myself as to whether I judged them without prejudice, or if I were being prejudiced because the criminals were responsible for the extermination of members of my own family. I knew at that time that only one of my siblings, a brother, was left alive. There was but one of my eight in-laws remaining and only four of their forty offspring were saved.

I tried to divest myself of all prejudices when I had to decide whom to arrest and whom to strike off the list of the naturally-prejudiced Hungarian government. But, I admit that it was hard to be just. When one is confronted by people who were at best neutral in watching the extermination of 600,000 innocent people, it is hard to extend the benefit of doubt!

Knowing that any individual returned to Hungary would be dealt with quickly and harshly, I had to ask myself with almost each case whether I was just or whether my bitterness was carrying me away. Often, I spent a whole sleepless night trying to decide one man's fate. I knew that no one would share my responsibility and that no one ever would question my decision except the hardest taskmaster of all, my own conscience.

While I spent my days interrogating the prisoners and my nights studying their state papers, I nevertheless had to devote some time to the desperate search for members of my family. Tens of thousands of unfortunate ex-slaves and prisoners came to our office, and I questioned many about my relatives. Some of the younger Jews were put to work in and around the horror camps and exterminated only after starvation and torture made them unfit for work. Some of them lived to see liberation.

Once I received news from my nephew, Andrew Fay Fisher, in the United States that two of his nephews, my grandnephews, were alive in a camp near Wels about one hundred kilometers from Salzburg. When I rushed there with some food and clothing, I also found the husband of one of my exterminated nieces in the same camp. My grandnephews and the husband of my niece were on their way to Hungary within a few days, but one of the young boys could not stand the burning anti-Semitism in Hungary. He returned within a few days to a displaced persons camp in Germany and tried to escape to any civilized part of the world. He remained at the camp with 800,000 others for the doors of the civilized countries were shut.

Whenever I visited one of those displaced persons camps in the discharge of my duties, I always was reminded of an operetta that I saw in my childhood. The operetta was called—I believe—*The Grand Duchess of Gérolstein*.[3] In one of the scenes, a count is about to drown in a lake, and people are running around madly singing, "the count will sink, the count will drown." But no one goes to help the count, and the count finally sinks.

The whole civilized world was sorry for the unfortunate, homeless 800,000 human beings, and international conferences were held about them. But the doors of civilized countries remained closed. When, finally, Venezuela, Canada, or some other country decided that some of the homeless could be permitted to enter—no Jews were wanted. Some members of the Jewish Joint,[4] as I understand, did provide false papers to a handful of Jews in France and thus enabled them to pose as Catholics and gain admittance to some Latin American countries. What a parody of civilization! The Jewish Joint was proving that other Jews had to become Catholics in order to save their lives!

Acting on the order of General Donovan, I started to interrogate the beasts from Hungary. I began with Ferenc Szálasi,[5] the Hungarian führer who seized power when Horthy ordered his army to surrender. Szálasi ruled Hungary for four months, the darkest and bloodiest four months of the thousand-year history of the unfortunate Hungary.

It was a revelation to see Szálasi and, to this day, I cannot see what sort of group or mob could ever have accepted him as their leader. The only explanation I can find is that the mass killing and the mass plunder of the previous months had debased the people to unprecedented depth and made them ready to accept the rule of this man. He was certainly of very limited intelligence with the mind of a fourteen- year-old child. Without doubt, he also was insane. But, the last forty years had been insane years in almost the whole of Europe. Szálasi's insanity was proved by the decision forced on his "Crown Council" held with his Ministers at Attersee, Austria twenty days before the final German surrender. Szálasi's decision assured Japan that his government would not make a separate peace with the Western Allies.

Szálasi was not sorry for the misery he caused to his nation, and he arrogantly assured me that his ideology would be victorious in Europe, if not in the whole world. His arrogance vanished on the day that he was assured that he would be returned to Hungary for his punishment.

I spent weeks in the jail of Landesgericht in Salzburg to interrogate the most vicious and the most foul characters. It was not a pleasant job. Some knew me from the days of their glories. István Antal, who tried to subsidize the Hungarian press of the United States six years before, remembered me well. He began his interrogation by pretending insanity.

Ex-Prime Minister Imrédy knew me, and his greatest worry was to prove that he was libeled by his political foes in Hungary who charged that he had Jewish blood in him. Indeed, one of his grandfathers was a Jew, but Imrédy carried three pounds of German documents to prove that his father was a bastard, by a Gentile, and hence Imrédy was of pure Aryan blood. I found proofs in the tons of documents I had to wade through that Hungary's Minister of Finance, Lajos Remenyi-Schneller, received 30,000

Marks monthly from Hitler years before Hungary embraced Nazism. I also found proof that other Hungarian statesmen and high-ranking army officers had been traitors to Hungary by accepting steady bribes from the foreign German ruler.

I had the doubtful pleasure of interrogating the degenerate László Endre who was the foremost demi-god advocating anti-Semitism years before Hitler became a power. Endre was in direct charge of deporting 600,000 Jews to the gas chambers. He did this with so much sadism that the merciful government of Hitler had to object. I saw the notes of the Germans addressed to the Hungarian government (still under Horthy's rule) that pointed out that while the deportation of the Jews was a necessity, it was neither necessary, nor wise, to do it with the brutality of the Hungarian gendarmerie under Endre's orders. Yes, the German people were civilized people!

Being only human, I do not wish to deny that I looked forward to these interrogations with great interest, if not actual pleasure. But all the pleasure and all the glory went out of it when I saw the cringing beasts before me. There was not one in the forty whom I interrogated personally who was man enough to admit the guilt and stand on his feet. Without exception, all of them denied participation. When the beasts were confronted with official proofs, with their own signatures on the orders, and when they had to admit to lying, their stereotype defense was that they were forced to occupy responsible posts, and that they did what they did only to prevent someone worse than they from taking their posts. Of course, they were still lying, for I also had proofs that they were fighting for cabinet posts during the bloodiest and most infamous days of Miklós Horthy.

Miklós Horthy himself was not my guest. However, I found diplomatic notes from the Papal Nuncio[6] that told Horthy plainly that he was lying when he had tried to maintain the fiction that the deported Jews were taken to work in Germany and that their lives were not in danger.

None of the Landesgericht prisoners had the guts to defend their actions. They were whining like the yellow curs that they were, begging me not to return them to Hungary but to grant them trial by the "fair American courts." When I reminded them about their foul statements about the "Jewish democracy," the "plutocratic brigands," and such other complimentary statements about America during the war, they blandly said that "it was just war propaganda."

Each of such interrogations made me sick, turned my stomach. Instead of deriving the smallest satisfaction for being able to deal with the murderers of my relatives, I was unhappy at their very sight. I felt as if I would be contaminated by lepers in their presence, and I was very happy when Nuremberg decided to return all of them to their native country. It was meant that I cease to interrogate them, and I did not have to look anymore at the scum of the ages.

When reading the papers of the beasts, I felt that I was in the midst of the bloody-handed pirates of a by-gone age. Most of them had "the goods" in writing on most of the others, the information that documented receiving pay from the Germans and stealing from the plunder taken from the Jews of Hungary. The whole gang was rotten to the core, and I wondered what sort of a nation Hungary became to tolerate such a gang of thieves and gangsters at the head of the nation.

The Hungarian army was not one iota better than the statesmen. Generals, colonels, and other officers came to see me to tell about the plunder and the graft of the others. The whole Hungarian army skipped the battlefield, and the clothing, food, and treasury of the quartermasters were stolen by these heroes. "Affairs of honor" were daily occurrences in their camps, such as when they quarreled over a piece of spoil or when the wife of one officer charged the wife of the other with having stolen "four Jewish fur coats" instead of the one permitted by the higher authorities. Each officer charged the other with having taken more than "their just shares" of rugs, silver, or money confiscated from the Jews or from the quartermaster's depot.

As a youngster, I read Emilé Zola's book about war, *The Debacle*.[7] I took it for granted that writers exaggerate or "color" at best, and could not conceive demoralization as described in the famous book. In Hungary, I was confronted not by demoralization but by the utter depravation of a nation, the ultimate result of any long dictatorship. The deterioration, of course, started at the very top. Ex-Regent Horthy attempted to wash his hands like Pilate, and he blamed all the cruelties and crimes on the Germans and on his traitorous entourage. However, I had in my possession a piece of paper with Horthy's handwriting, dated 1938, which ordered the then Prime Minister Imrédy to go slow with land reforms and to expel most of the Jews from Hungary. I saw petitions from the deceased Prince Primate

Seredy of Hungary addressed to Prime Minister Döme Sztójay[8] (hanged later) that did not object against the humiliation of the Jews by forcing only them to wear the Jewish Star[9] but that requested exceptions for the converted Jews.

I have in my possession perhaps the most infamous paper in the world, a paper written by the secretary of a Catholic Bishop. The paper contains the Bishop's plea to Sztójay to limit the brutalities in connection with the deportations and reminds Sztójay—in the name and at the order of his Bishop—that "after the war we will be taken to task for the method of the deportations, but should the Anglo-Saxons win, we may even have to answer for the fact itself." His Eminence was not worried about the "fact" of the deportations, except in case of an Anglo-Saxon victory, only about the method of same!

The return of the Hungarian war criminals, all of them, was arranged between General Eisenhower, Justice Jackson, and General Key of Budapest who was head of the American Mission in Hungary. The Russian High Command of Hungary and General Key insisted that Hungarian guards should take the criminals home. But since the criminals were in the custody of our unit, we wanted to deliver them to the American Military Mission of Budapest ourselves. The Russians were unwilling at first to grant clearances for American guards, but the Hungarian public was clamoring for the return of the criminals who were the cause of the nation's downfall. Finally, the clearance came through.

Since I was in direct charge of the interrogation activities and the criminals, I was included in the group that was cleared. I thought that finally, a year after I was alerted to go to Budapest, the door was open. I was mistaken.

The Allies had proposed a condition that the Russians and Hungarians were unwilling to meet. The proposed condition stated that the courts of Hungary were to be free to sentence any of the criminals but not to execute them without special permissions from Nuremberg, a very reasonable precaution before the trials of the Germans. Some of the criminals perhaps would be needed at Nuremberg as a witness prior to their executions.

The negotiations took weeks of delays, and I was to leave Europe by the end of September. I was sick of the experiences of the last few months, my job was done, and I begged to be released at the end of September. My request was granted.

I still wanted to see my last remaining brother, and I was fearful that the negotiations would not end in time for me to make the trip. A few days before the end of September, I received a letter from the then Hungarian Prime Minister and temporary head of the state, Béla Miklós.[10] He praised my services to Hungary and invited me to Budapest as the guest of the government. Since I was cleared previously by the High Command of the Russians in Hungary, and in the face of the letter of the current head of the government, the Russian commanding officer of Vienna had no choice but to grant me the necessary road pass to Budapest.

CHAPTER 20

In Hungary

★

I LOADED MY STATION WAGON with medical supplies until it sagged, and I had to remove part of the load. One of my most loyal assistants, a Hungarian paratrooper, and I then were on our way to Hungary and Budapest.

We arrived late in the afternoon, but not too late to see the ugly ruins of Buda,[1] the complete destruction of the famous King's Palace on top of the Buda Mountain,[2] and the sad sights of the destroyed bridges of the Danube. It was too late to present myself at the office of Prime Minister Béla Miklós, and I accepted the overnight invitation of one of my friends who was liberated in a German death camp by the Americans. The paratrooper and I were my friend's guests for the first night. I was not quite up to calling on my brother and learning the final fate of members of my family.

We slept in windowless rooms, and it was cold, but there were very few windows left in Budapest after the siege of six weeks. I walked the streets that first night with mixed feelings. I could not help but feel sorry for the people. They were hungry and cold. But the people asked for their fate and then some!

Hungarians, of course, will keep on telling the world for at least a century that they had no choice, that they were forced to join Hitler and to participate in the unspeakable crimes of Hitler and the Hitlerites.

It is not so!

Before Count Pál Teleki,[3] the last decent Hungarian statesman of the Horthy Regime, killed himself rather than participate in the infamy of attacking Yugoslavia three weeks after an "eternal friendship" treaty was signed between Hungary and Yugoslavia, Teleki wrote a letter to his Minister at the Vatican. Teleki said in the letter that he blamed himself for accepting the "Vienna Award" from the dictators (Part of Transylvania was returned to Hungary by this award.), and he told about his struggles to keep Hungary out of the war. But—said Teleki—when Horthy was promised the sea by Hitler, the Admiral Horthy became "Fier and Flamme" (Fire and Flame), and there was no keeping him out. Even most of the present "Democratic" leaders of Hungary greeted the invasion of Yugoslavia with joy and expressed their admiration for Horthy in glowing terms.

When Germany declared war on Russia, Vyacheslav Molotov[4] begged the Hungarian Minister, László Bárdossy,[5] to take back his declaration of war on Russia and report to his government that there was no reason for a quarrel between their two countries. Hungarian Prime Minister László Bárdossy kept the Minister's wire from parliament until it was too late to change the course of events. And, of course, it was the most miserable ingratitude and the most insane delusion of grandeur for Hungary to declare war on the United States.

Following the first terrible impact of World War II and the loss of about 100,000 men, Hungarian Prime Minister Miklós Kállay[6] refused more troops to the front, and Hungary enjoyed the prosperity of war for years without increasing the sacrifices. And the nation was satisfied. Proud as the Hungarians used to be, they reconciled themselves, without much struggle, to the German occupation when Horthy was trying to surrender and was kicked off for having been bribed by the plunder of the exterminated Jews.

These are not nice things for me to say about the nation and race to which I once belonged, but far better that this should be said by a man of Hungarian birth who also feels and is able to measure the mitigating circumstances. When I indict the Hungarian nation for embracing Nazism without much struggle, I have to hold responsible the worthless lesser-gentry class, the class that obtained power—with Horthy's full approval— in the 1930s. And when I indict the nation for the extermination and plunder of 600,000 Hungarian Jews, I have to give credit to about 200,000 Hungarians who risked their lives in hiding and saving some of the Jews.

But no matter where the responsibility or the guilt was, the masses were suffering beyond description for Hungary was now as bare as a desert. Hungary had been cleaned of grain, livestock, industrial equipment, and tools of every description. On my way to Hungary from Vienna, I saw peasants plowing with a calf, with one horse, or with two men pulling the plow. I also saw men trying to break the ground by hand with a hoe.

On my first morning in Budapest, I went to see my brother, Jakab, and I found him in pitiful condition, although he was far better off than many of the Jews who escaped. His wife, son, daughter, daughter-in-law, and grandchild were saved. The only loss in his immediate family was his son-in-law. He and his wife were chased from one ghetto to another, but his daughter was kept safe in an insane asylum by the chief of the institution who kept scores of Jewish people at the asylum. His daughter-in-law was hidden in a basement for months by one of his loyal employees. His son returned sick with fever from a German camp. Even my brother's shop was intact upon his return; tools and lumber did not appeal to the plunderers who chose instead jewels, furs, coats, silver, and rugs.

Although my brother owned a four-story house, he was homeless. His home consisted of five nice front rooms on the second floor and a kitchen. The front of his house was hit by a bomb, and there was no outside wall left to the five rooms. The rooms were open, the ceiling was supported by props, and the family of six, as well as two homeless relatives, lived in the kitchen. Some of them slept in the open rooms during the summer, but by the end of September, it was too cold to sleep in open rooms. They then had to pile up as best as they could in the small kitchen.

I had Jakab contract immediately for the re-building of his home. At least I was able to help my one remaining brother. I also provided a small capital for my two nieces and my two grand-nephews. My brother reminded me, with tears in his eyes, of the time that he refused to take the same coach with me when I first left Mátraverebély for America.

Hungary accepted the terms of the Allies regarding the war criminals within a very few days, and this was to the great satisfaction of all of Hungary. I was still in Budapest when the first shipment of criminals arrived. However, I was bewildered by the changes in the Hungarians.

The corruption in Romania was a by-word in Hungary for decades as well as anywhere else in Europe. Before World War I, public life in Hungary was as clean as in any country in Western Europe. But nothing that ever existed in Romania could have been one-tenth as bad as that which I now saw in Hungary. In fact, Hungarian Prime Minister Béla Miklós told me at our first meeting that invasions and destructions were terrible, but far worse was the corruption and the destruction of public and personal morale. Two days later, I learned that the Prime Minister himself had accepted a vineyard confiscated from a Jewish owner and although by a miracle the original owner returned, the Prime Minister refused to return the vineyard to the rightful owner.

Beginning on the second day of our stay in Hungary, the paratrooper and I were guests of Colonel George Kovach of the Military Mission. He gave a party on the first night of our stay in his villa. Not one of the ninety guests—the elite of Budapest—dared to leave Kovach's villa before daylight for hundreds of Russian deserters were roaming the streets of Budapest and undressing civilians in the darkness of the evenings to fulfill the Russians' desperate need for civilian clothes.

There was, and there still is, a burning anti-Semitism in Hungary, the like of which was unknown even in the days of the deportations. I was surprised that in the face of all the sufferings of those who escaped, and in the face of 600,000 martyrs, Hungary was yet unreconciled and still claiming that only the bad Germans were to be blamed for the persecution, plunder, and extermination of the Jews.

I went to Mátraverebély hoping that its population and my old schoolmates who were so very grateful for my assistance extended to them during World War I would have saved something of the personal property of my widowed sister. Since they failed to save her, I hoped that I could turn over to my nieces whatever was left of my widowed sister's property. My old friends were somewhat embarrassed at first, but soon warmed up. They came to see me by the dozens, hoping that history would repeat itself and American dollars would be forthcoming once more. But, when I inquired about the personal properties of my late sister, no one knew anything about anything. Her immediate neighbors blamed everything on the Russians. I said that if anything were taken by the Russians, it was all right because we are Allies, and I would recover everything. But if they were libeling the Russians without cause, it would be too bad. The story was then immediately changed. Not the Russians, but the gypsies, were the robbers. To make a long story short, not even a match stick was recovered in the village where I was so popular even seven years before. The same story could be repeated 600,000 times in Hungary.

My Gentile friends—everybody was a friend to any American in uniform—explained the reasons for the intense anti-Semitism. The real reason, of course, was the fact that the witnesses and victims of the horrible crimes returned to remind Hungary and the individuals of the foul deeds.

The witnesses and victims not only returned, as many did return, but perforce demanded their old positions if they had held public or private offices before the passing of the Jewish laws. Storekeepers likewise demanded their stores, property owners their properties, and tenants their living quarters. And professionals, businessmen, etc., offered competition once more.

The Hungarian government, of course, re-instated every one into all their rights, and real and personal properties had to be returned—on paper. The government was powerless to enforce the newly-passed laws and decrees and, while upon request, the government furnished police to assist in the recovery of plundered properties, the government was helpless in the face of opposition of the thieves.

The government in Budapest was more or less able to enforce justice. But, if a Jew in a provincial town insisted on the return of his stolen possessions, he simply was chased out of the town once more, if not lynched. And I am sorry to say that the so-called middle class is equally guilty with the peasants. While the murderers and plunderers once did their dark deeds under the black flag of Nazism during World War II, they continued the murdering and plundering under the red flag of Communism.

The Hungarian Communist Party tried to impress Moscow with its popularity, and its leaders invited the so-called "small Nazis" into the fold with the consequence that at least 70 percent of the members of the Hungarian Communist Party were recruited from known former Nazis. Just as the thieves and plunderers were protected during the Nazi Regime, they now enjoyed the protection of the Communist Party, and there were several mob actions and lynchings without serious consequences. Once a person has carried the Communist membership ticket, he is all but untouchable. It is of public knowledge that several ex-Nazis who were afraid to return to Hungary and answer for their crimes received their Communist Party ticket while in Germany and Austria and thus returned home fully protected.

Unfortunately, most of the Moscow-trained post-World War II leaders of Hungary were Jews, as were the leading figures of the Béla Kun Regime in 1919. This fact tends to turn even the most decent Hungarians against the Jews.

The post-World War II Hungarian Communist Party contained good, patriotic Hungarians who opposed the subservience of the party to Russian interest, and the party also contained as many ex-Nazis. The Hungarian Communist Party was thus divided, though the division wasn't visible for the moment. There was a strong nationalistic and anti-Semitic wing under the leadership of the Minister of Interior—whose several brothers held high offices in the Nazi Regime—ready to take over if and when the Jewish leaders had done the "dirty work" of the party.

I left Hungary in disgust. I decided once more, as I did in 1938, that I never would return there.[7]

CHAPTER 21

Return Home

★

IN OCTOBER, 1945, I returned to Washington with the intention of terminating my services. I, in fact, did leave the OSS. The OSS was the most unique organization America ever had, and I will remain ever grateful for the privilege of having served in it for those years were the most interesting years of my life.

There was, and to some extent there still is, a misconception about the OSS organization in America and even more so in foreign countries. Necessarily, the OSS had to be a secret organization. As a result of this necessity, the OSS became the target of radio commentators and newspapermen who were ever eager to reveal secrets to the public.

The OSS became known as a spy organization, and the public was not aware that the OSS did many other things besides the necessary intelligence work. In fact, the OSS had six or seven different branches to undertake some of the most hazardous tasks of World War II, and only one branch concerned itself with intelligence work. Following World War II, General Donovan had some films made, and many articles and several books were published about the OSS. Perhaps our own public then changed its attitude toward the OSS, which went out of business with the close of World War II. But in foreign countries, OSS ex-members will be forever marked as spies. My duties did not include intelligence work, although Mr. Rakosi, the Communist Chief of Hungary, did not fail to refer to me as an intelligence officer at our first meeting.

Long before any member of the OSS left America, the Germans were well informed by the newspapers about the alleged tasks of such men. As a result of the needless publicity, the Russians believed that every ex-OSS man must be kept under permanent surveillance in the occupational zones. I wonder if the newspaper publishers ever thought about the consequences of their unfriendly remarks?

Although the OSS was built in a hurry, I believe it functioned remarkably well and that it was the crystallization of American democracy. Starting at the first training school where colonels, captains, enlisted men, naval officers, and civilians had to appear in plain Private GI clothes, the democratic spirit of the organization was maintained everywhere. The OSS gathered approximately twelve thousand men in a hurry for the various tasks undertaken by the organization had to reach to all strata of society for propaganda, organization work, demolition, intelligence, etc. If a few unfit men crashed the OSS service gate, it is no wonder that so few of them succeeded. No one knew in advance the dangerous assignment for which he was recruited, and the cream of the men volunteered from the Navy, the Army, and civilian life.

At one time, American media charged that society folks were invading the OSS, and we did have a few snobs with whom to contend. We also had a few who sought shelter against draft, but such men were weeded out as fast as it was possible or just hated out by the vast majority of the decent members. As far as my OSS experience was concerned, I, for the first time in my life, did not feel that I was an outsider. I belonged, and my ex-comrades will never know how grateful I always will be for their friendship, for their cooperation, and for everything that they made the OSS to be for me.

Johnny Bako died while I was in Italy and with him my hopes of a contented life on our farm also died. Between the two of us he was to be the farmer, and I was to earn money by writing for several Hungarian newspapers.

I was told in Washington that I would have to return to Europe as soon as I had attended to my personal affairs, and I went to California to dispose of our farm. I became homeless once more, as so often in previous years, and eight weeks later I returned to Vienna. I was not happy about the extension of my years in service. I felt that I was getting old and ought to find a permanent home as well as a new plan for my declining years.

I knew that I never would enter the foreign publishing field again for, as a publisher, I'd had great doubts during the last ten years as to whether foreign newspapers truly rendered a service to the immigrants. The newspapers, churches, and societies of the immigrants' own languages are a great convenience and are of great assistance to the immigrants during their first two or three years in America. And priests, ministers, and editors stay at their posts unselfishly in this belief, although most of them are talented men. But, after publishing Hungarian newspapers for fifteen to twenty years, I realized that by the mere fact of having newspapers, churches, and societies of their own, the immigrants were more or less isolating themselves from America and building, so to speak, their own ghettos.

When I visited cities and towns with only two or three Hungarian families, I was amazed to see that the families had become integral parts of their environments, part and parcel of the community in which they lived. The Hungarian peasant woman was just as active in organizing church socials with her native American friends—after a few years' stay in her community—as her sister in the Verhovay Segély Egylet[1] in Cleveland.

And just because I knew that I would have to choose another line of work, I was not eager to spend another half year in Europe. I was promised that I would be free to return to America within six months.

I did not like Austria, and I did not like Vienna. And I did not like the Austrians.

I had experienced the friendly, smiling sufferers of Italy, and then I was confronted with the sullen–faced Austrians, ever-whining about their misfortunes, denying the smallest bit of their responsibilities for their fate, and choosing once more the role they played so well after the First World War—the beggars of the world. Instead of the "Gemütlich Vienna," I found wooden-faced people with never a smile, with never a song. The Austrians were unwilling to get down to realities and to work, and they expected the world to provide for them.

My work still concerned me with Hungary. A tremendous part of the national wealth of Hungary was taken by the last Nazi government to Germany and Austria, and much of the wealth was now claimed by foreign owners. Dutch, Swiss, and British owners of Hungarian concerns were seldom willing to have their equipment returned to Hungary due to the possibility that the equipment perhaps would be forwarded on a reparations account to Russia. An immense amount of research work became necessary for me. America also was getting ready for the Peace Conference of Paris, and every branch of the Military Government was called upon for endless information and dates.

One of the most interesting problems that I refused to touch was the so-called "Gold Train," a trainload of confiscated Jewish property containing jewels, furs, objects of art, rugs, and other valuables. Most of the owners were liquidated in Auschwitz, but many returned, and they had a just claim on their personal properties that often represented small fortunes.

The Jewish Joint, or at least some representatives of the Joint, claimed the whole trainload for the benefit of international Jewry. This claim was justified on the grounds that the Joint organization spent millions of dollars for the benefit of Hungarian Jews who came back and also on the further grounds that many of the owners still alive were scattered all over the world with the assistance of the Joint. The Hungarian government never tried to deny that the goods belonged solely to the Jews, but some members of the government held that if the proceeds were to be used for the benefit of the whole nation, it would perhaps lessen anti-Semitism. The Hungarian Jews themselves were divided, but most of them wanted to take a look and find their own properties.

Once, the Chief Rabbi of Budapest called on me with the request that I be the arbitrator for the recovered properties of the returning Jews. He complimented me with the statement that I had the confidence of the Hungarian Jews, as well as of the then government of the state. When I declined with thanks, he proposed an American Committee to oversee arbitration, and I assured him that no American in his right sense would ever touch the question with a twenty-foot pole. Truckloads of gold and suitcases full of diamonds were buried and recovered in French-occupied zones, but as far as I know, no one at present knows the whereabouts of this treasure.

Hungarians visiting Vienna or the American zone of Austria invariably called on me for assistance in getting the necessary permits, and I was sick to note the black market operations they engaged in. High government or city officials were just as eager to do a little black marketing, besides their official business, as anyone else, and I often was embarrassed by their misuses of the permits secured for them. I was amused at first. Later, I got mad to note the number of teaming Russian agents who mostly were Hungarians invading the American zones under various pretenses. Many of the invaders offered to furnish unwanted intelligence about the Russians, and I am sorry to say that American authorities were very lenient with the invaders, at worst sending them back to Hungary.

American journalists are still naïve enough to publish everything pertaining to the Military Government personnel, size of the troops, equipment, etc., and I did not see the need for our Eastern ally to send too many agents for additional information. I had a tough time convincing — but I did convince — some of the Russian agents offering to spy on Hungarians and Russians in Hungary that America was not interested in their spying, and that I personally was not connected with intelligence work. My OSS past remained with me during my stay in Europe.

To gather the necessary dates and information needed for the Paris Peace Conference,[2] and more particularly to clear titles of Hungarian assets in Austria, I had to make numerous trips to Budapest during 1946. I was astonished at the pace of the work of reconstruction in Budapest. Budapest workers had no equipment, such as steam shovels, trucks, etc., which the Germans and the Austrians had in abundance. The immense amount of debris in Budapest was cleaned with hand shovels in short order, and overnight tram cars took the debris to be dumped outside the city limits. Hungary's railroads were in fair shape within a year despite the fact that all railroad bridges, large and small, had been destroyed. Many of the damaged or destroyed buildings of Budapest also were repaired or rebuilt within the year.

Hungarian factories were re-opened, with or without some equipment, production started, and foreigners were amazed at the strength and the vitality of the Hungarian nation, a strength and vitality unparalleled anywhere in Europe. The hard-working people of the unfortunate Hungary certainly deserved a better fate and a brighter future than Hungary's present outlook.

An old lady came to see me in the Hotel Bristol on one of my 1946 trips to Budapest. It was not unusual to have five, six, or ten callers before breakfast for, after breakfast, I left the hotel and seldom returned before 10:00 P.M. The old lady was dressed in black, her hair was gray, and she sat patiently in the lobby, waiting to be last amongst the morning callers to see me. When she introduced herself, I was stunned.

She was Icza!

Her husband, an army officer, was somewhere in Austria or Germany, and she begged me to find him. We did not talk of old times; perhaps both of us were afraid to open wounds healed long ago. But, she told me about her three children, and she invited me to her home. I had to decline the invitation. I put the machinery in motion to find her husband upon my return to Vienna. Two weeks later, I learned that he was dead. I did not have the heart to forward the information to Icza.

My work took me to Paris several times during the Peace Conference. I happened to be there at the time of Mr. Henry Wallace's famous statement. I never can forget the utter chaos caused by his bombshell. The whole of Europe was reeling from his punch, with the exception of the Russians. Paris became a madhouse for a few days, and if Mr. Wallace had made his appearance in those days, Americans would have dealt with him very harshly.

It was heart-breaking to see the American delegates left out on a limb, even by America's President, and we felt that the work of every one of us was futile as long as an outstanding American, a member of the President's Cabinet, was ready to knife the American delegation in its back. Wallace's stand must have convinced the Russians that the people of the United States were spilt in two equal camps, and much of their bellicose attitude may be credited to Mr. Wallace.

When in Hungary, I saw the ravages of the terrific inflation and the equally serious consequences of the drastic deflation, both managed by the Hungarian-born famous Soviet economist, Jeno Varga. The inflation and the government-fixed prices drove all industrial concerns into near bankruptcy and forced them to borrow more and more of the state-issued currency every week, until every one of them became heavily indebted to and at the mercy of the Communist-managed Supreme Council of Economy. I could not help but see that the fixing of the unreasonable rate of exchange of the new currency would result in the complete isolation of Hungary from the Western World. Every foreigner left Budapest within a week of the stabilization for a small room in the Hotel Bristol cost anywhere from $12.00-$18.00 per night, and the foreign buyers of goods could not see their way of meeting Hungarian prices.

The same Mr. Varga was loaned later by the Soviets to manage the inflation and the future stabilization of the Romanian currencies, and he is the real author of the respective three, four, and five-year plans for the various Balkan countries. The financial plans of Hungary, Yugoslavia, Czechoslovakia, and the future plans of Romania dovetailed with one another, and all of them would be harmonized with the master plan for the Union of Soviet Socialist Republics.

The Hungarian government's price-fixing policy set the prices of industrial products and consumers' goods very high and the price of agricultural products very low. This was a clear plan to liquidate the "kulah" economic class that was comprised of individuals still having fifty or one hundred acres of land in Hungary. A quintal[3] of wheat, in pre-World War II Hungarian days, was worth a pair of boots; in post-war days, two and one-half quintals were required to buy the cheapest pair of shoes, and almost a quintal was required to buy the cheapest shirt. Almost three quintals were required to buy a good shirt. The wages of the agricultural workers always had been too low in Hungary. Following World War II, the agricultural workers' wages were equal to the highest-paid skilled workers, and no land owner could make both ends meet if he had to hire workers.

When I dealt with the financial affairs of Hungary, I was up against the same problems that I had to face when gathering up their war criminals. When I considered all of Hungary's past mistakes, or crimes, if you please, for which they suffered plenty, I deeply sympathized with them in their heroic struggles. And I had to search my soul from day to day to be certain that I was not carried away by my sympathy and that I was thinking and acting with the absolute neutrality expected of me in my work.

Luckily, the policy of America and our government was, and still is, to assist all European countries. My superiors were always in harmony with my outlook on one hand while the Hungarians were appreciative of my efforts to assist them, especially in regaining their displaced properties, on the other hand. As hard as I tried to convince the Hungarians—from the President of the Republic down to ministerial officials—that my attitude was solely governed by the policy of my government, they attributed much of my attitude to my Hungarian birth which often was embarrassing to me.

Deep in my heart I knew, and still know, that my Hungarian birth had nothing to do with my activities, for I have even more sympathy for the equally hard-working people of Italy. Nevertheless, it was good to feel that my work was satisfactory to both parties, except that the satisfaction of my superiors kept me in Europe much longer than originally planned in Washington.

Prior to my leaving Europe and the service, I had to go for a short visit to the Middle East, Egypt, and Palestine, and I was astonished to note the low esteem that the Arab world had for the British Empire. While I was in Cairo, a British High Commission arrived there to discuss the possibility of scaling down British indebtedness to Egypt, but the Commission failed even to meet any Egyptian authorities to debate the question. The Commission was just told that Egypt had nothing to discuss.

Nationalistic feeling in the Middle East was high. Europeans were discharged from responsible positions, business houses maintained by Europeans almost were boycotted, and visas for European businessmen were refused. Natives in high and low circles were gloatingly discussing the end of the British Empire. Whether our British cousins would be able to keep India, or a part of India in the Commonwealth, or not, the power and prestige of the British in the Middle East was definitely over. And, clinging desperately to the Sudan would not help the British. Although all white people are subjected to this Middle East antagonism, the Middle Easterners still have a soft spot in their hearts for America and the Americans.

Originally, I went from Rome to Palestine via Cairo with a transit visa to Egypt for a single entry. I neglected to secure another transit visa for the return trip, and a young Egyptian officer at the Cairo airport had me wait in his office until he finished with the rest of the passengers.

Two minutes after I went to his office, a servant came with a cup of Turkish coffee and cigarettes. When the officer joined me after ten minutes, he apologized for the delay. He explained that while an emergency visa would be granted—for a stay of several weeks, if desired—he was not authorized to issue same without first reporting the case to his superior. The visa was affixed to my passport within another five minutes and without the officer's knowing that I was on official business. My face was red as I thought of the possible method and treatment in New York if our situation had been reversed. Instead of receiving apologies, coffee, and cigarettes in New York, my young Egyptian friend would receive a far less friendly treatment—even with proper visa—at the La Guardia Field.

I do not know how long Egyptians will treat Americans with the courtesy shown to me because the other passengers and employees of the Transcontinental & Western Air had become impatient. A uniformed TWA man came to the office during my interview and asked me the unforgivable rude question, "What do the Wags want of you?", although the TWA man knew that the Egyptian officer spoke English.

The "Wag" term irritated my Egyptian host plenty, and my host told the TWA man to get his bus off the field because my transportation would be taken care of by my host. My host refused to accept a package of Camels until I took his package of Players and, within minutes, we parted as friends. I rode to Cairo in a limousine. I am afraid that if America is to maintain our prestige and popularity in the Middle East, we have to leave the words "Wags" and "natives" out of our American vocabulary.

My work was immensely interesting, and I derived much joy in doing it. I was associated with the most wonderful group of people and the finest bunch of friends that any man ever could hope to have. My chief and good friend was the finest America can ever produce, and he selected his co-workers with such a care that we became the outstanding small unit of the American zone in Austria. I knew that once I left my friends, I could never hope to find such congenial people and so many truly good friends.

But I had had enough of Europe, and my personal affairs also urged me to return home. For one who is not thinking of the advantages of the black market value of the dollar, for one who is not seeking to gorge on the misfortune of the people, for one who can see and hear and is not blind to the miseries of the multitudes, Europe is very depressing. Apart from going around amongst hungry, naked, and freezing people, a hard task by itself, the future of Europe seemed so dark that it scared me.

The countries behind the Iron Curtain were sentenced to be swallowed up by a government system that they abhorred. The countries were helpless to offer any resistance, and one could feel the tremors of future bloody eruptions.

The people of Western Europe, including Italy, were bewildered and were desperately waiting for — they did not know what. But, whatever it would be, it would be terrible. Subjected to the merciless inroads of the new faith, most of Western Europe couldn't decide whether they were afraid or hopeful of a new world war, a new clash between East and West. Left to their own fates, without a final accounting between East and West, Western Europeans felt that civil wars were inevitable. They were afraid that Europe would sink once more into the abyss of a new Dark Age.

Wherever I went in Europe, I saw the masses in desperate doubt as to how and where to seek succor and salvation. Their instincts and sympathies were for the Western civilization and style of life, but they were somewhat bedazzled by the threats and promises of the so-called "people's democracy." And, while they longingly looked toward America, the far-away beacon of light, the kind-hearted giver of untold alms, they listened day to day, hour to hour, to the preachers of hatred who disseminated unbelievable lies, falsehoods, half-truths, and vicious libels about the only hope for mankind — America.

And America was all but silent.

I was besieged for information and explanations wherever I went. I had to answer millions of questions regarding the various statements of Moscow and her hirelings, and Congress reduced the pitiful small sum requested by our State Department for Information Service.

I saw the governing classes, the elites of the European nations, cynically resigning themselves to the inevitable clash either between West and East or between the two factions of their own nations. I saw the governing classes turning into hedonists, trying to squeeze some enjoyments out of every day instead of getting down to business and leading their people into light.

The utter demoralization and the disintegration made me sick to my bones, and I wanted to spend as much of my declining years as possible in the clean and fresh air of America. I was paid a very decent salary, but very little was saved, for Rockefeller's fortune would not be enough to help each deserving unfortunate with whom, on my job, I came in contact.

My own very uncertain future and security were urging me to return to America, and, after several postponements, I was permitted to leave Europe and the service in June, 1946. The homecoming was not quite as happy as the anticipation, but it seldom is.

All my fears about the future of Europe were about to materialize, unless Europe were given American help to the utmost ability of our nation. I found the American people in the throes of doubts and our American politicians in the midst of the presidential campaign. While the shadow over Europe grew darker and darker, America still doubted its own future course to action which might once more lead us to the tragedy of "too little and too late." And the darkness under the shadow could develop into a new Dark Age.

My own little community, Hungarian America, was no better off. The radical group was more vocal than ever and tried to sell the new Red future of Hungary to Hungarian-born Americans who, of course, would always think of their native land with the understandable nostalgia and sympathy as they were stampeded into the Red sector of American life. The other extremists, the erstwhile Nazis, were equally loud. They felt that their previous stand was justified for, lo — Hungary was about to be swallowed up by the Eastern giant, and the Nazi venom was permissible now that America was at odds with Russia.[4]

Hungarian Americans were beset by doubts for, loyal as they were to America and everything America meant to them, they were worrying about the fate of their native land, too. I had no desire to return to newspaper work or public life in one of America's Hungarian colonies, and I had little reason to change my mind when I returned home.

APPENDIX

Two Men, One Evil

The Washington Post
December 16, 2003, Tuesday
Final Edition
Byline: Charles Fenyvesi
Section: Editorial; A37
Length: 919 words

The capture of Saddam Hussein frees me from a promise of secrecy I made to a retired colonel of the Office of Strategic Services; the precursor to the CIA, in 1957. I was in my first year of college, and an old family friend called me to say that a friend of his who also knew my family was at his house for a visit and wanted to talk to me.

I knew the visitor by reputation: retired OSS officer Martin Himler, publisher and editor in chief of a Hungarian American newspaper who was recruited to work for the OSS during the war and became head of the Hungarian section. In occupied Germany it was his job to track down, arrest, interrogate and eventually return to Hungary suspected Hungarian war criminals.

In 1957 Himler was a bald, little old man in his sixties, but he was still muscular and quick on his feet. He had emigrated to the United States as a young man. He came from a Hungarian village not far from where my family lived. He was Jewish, and I remembered his name mentioned as one of several men who had been in love with my Aunt Elza, a red-haired, soft-spoken beauty whose life ended in Auschwitz.

I called him Uncle Martin, as young people in the old country are expected to call their elders. He asked about members of my family. He knew who survived World War II and who did not, and he asked only about those who survived.

He asked me whether I knew that he had interrogated Ferenc Szalasi, head of the Arrow Cross, Hungary's equivalent of the Nazi Party, and the head of state for several months after the Nazi putsch of October 1944. I said yes and added that I also knew about the U.S. award he received for his interrogations of 71 Hungarian Nazi leaders in the summer of 1945 in Austria.

l passed a test. "There is something l need to tell you," he said, and he took the long breath of a man ready to unburden himself of a secret. "But could you keep it to yourself?" I said yes.

Himler told me that he had carefully arranged the setting for his interrogations. A U.S. soldier escorted each prisoner to his office, saluted him and then asked for permission to leave the room. He said yes and told the soldier to stand outside the door and let no one in.

He sat at his desk, in uniform, of course. He conducted the interviews in Hungarian. The prisoner had to stand. One after the other they said that they would tell everything but that they were innocent. All the decisions about the deportations of Jews and other forms of cooperation with Germany depended on Szalasi, the fuehrer, called "the leader of the nation" in Hungarian.

Himler told me that he was stem and impersonal, suppressing all emotion. His job was to gather information and report back to the OSS. He had completed at least a dozen interviews before he decided to have Szalasi brought into his office.

Himler said he barely looked at the man when asking the basic questions: name, place and year of birth. Szalasi answered, then he said that as a head of state, he was not required to answer any further questions.

Ignoring the comment, Himler asked the next question: "What is your profession?" Szalasi said: "The leader of the nation."

"Suddenly I lost my self-control," Himler told me. "l got up from my chair, walked over to Szalasi, who stood ramrod-straight, and l slapped him on the face so hard that he staggered and almost fell. I walked back to my desk and sat down. After Szalasi straightened himself up, he said that he was ready to answer all my questions."

Himler looked at me and said: "To this day l am ashamed of what I did. As an American officer I shouldn't have resorted to violence with a prisoner of war." Himler stared at the carpet in front of him. "Don't tell anyone

APPENDIX

Newspaper clipping showing Himler's editorial on the front page of the *Hungarian Miners' Journal*, January 2, 1958. The Editor's Note mentions a reprint in the *Washington Post*, which ran on January 15, 1958.

what I have done," he said, "but knowing your family, I wanted to share my secret with you."

"Uncle Martin," I burst out, "you slapped that murderer for my family too, for those who perished like my Aunt Elza and those who survived. You have nothing to be ashamed of."

Himler got up and kissed me on both cheeks. "Thank you," he said.

Himler died many years ago. His memoirs do not mention a word about the slap. Nor does he mention my beautiful aunt.

Szalasi was hanged in Budapest. Nearly a half-century after Himler told his story, l am watching video images of American soldiers checking Saddam Hussein for lice and taking some of his saliva to collect a DNA sample. Soldiers of a democratic nation captured a tyrant and a mass murderer, and gave him the kind of humane treatment he and his henchmen never accorded to those who disagreed with him. Two sets of standards, two kinds of humanity. But l will not criticize Himier for his loss of self-control, for his momentary lapse of appropriate conduct. I think he had nothing to be ashamed of, and I think that after 46 years his secret can be told. I admire the rules of civilized behavior, but victims and others outraged by mass murder cannot always hold back.

I watch Saddam Hussein on television, but the men I see in my mind's eye are Szalasi, Hitler and Stalin.

If I Were The President...
BY MARTIN HIMLER
An Editorial in the Hungarian Miners Journal, Jan. 2, 1958

If I were the President of these United States I would realize that it is indeed time for greatness, and would act accordingly.

I would subject the Kremlin to an acid test, by submitting to the world a bold, a daring plan, a plan that would re-establish America as the shining hope of the suffering world, as she once was.

A plan that would be as great as our United States.

I would call a summit meeting without much delay, and with previous approval of our allies and of Congress I would submit — openly, in plain sight of the whole world — the following plan:

Both sides to put an immediate stop to testing atomic and hydrogen bombs.

The NATO nations to evacuate West Germany, Russia to evacuate East Germany.

The Americans to abandon their overseas air fields, Russia to evacuate the satellites.

Both nations to cut the armed forces and defense budget in half, with the British and French, in fact all other nations to follow suit.

The immediate abandon of NATO and SEATO, also the WARSAW alliance.

Recognition of the Chinese government, and dropping Chiang-Kai-Shek by the U. S. with the Chinese government, recognizing the independence of Formosa.

All of these moves under strict international control, with the right of both parties to inspect each other's military and naval establishments.

With defense budget cut in half, *the United States is to contribute twenty billion dollars yearly, during the coming ten years, to an international fund, to be managed by the United Nations, to assist the newly liberated have-not nations.*

The USSR to contribute ten billion dollars yearly to the same fund, and a proportional contribution by every other nation profiting from the cut in their own defense budget.

With approximately forty billion dollars each year, the have-not nations would soon reach a tolerable level of prosperity, and the threat of general anarchy would cease everywhere.

And with the new peaceful atmosphere, and new prestige and power of the United Nations, such minor disturbance of Cyprus, Izrael, Jordan, etc. would soon be settled.

If the Kremlin is sincere in seeking a just peace they could hardly turn down the offer.

If they play up the everlasting prattle of peace only for propaganda purpose, let them beat, or even meet this offer.

Utopia?

It may be so.

But it is high time to take the initiative from the Kremlin, and to show the world the sincere goodwill of the United States.

Is it too high a price to pay for peace and for the sincere admiration of the peoples of the whole world?

What is the alternative?

Increasing tension between the two camps, increasing burden of armaments, forty billions for next year, more and more in succeeding years; until the burden would crush mankind in both camps; and perhaps an end too horrible to contemplate.

Let us consider our own position soberly.

We could stand forty billion dollars yearly, with twenty of them for home defense, and twenty for assisting our fellow men everywhere.

And every American would be far happier to spend half of our present defense budget for constructive charity, than for sterile and useless war machines.

With twenty billions to be contributed in machinery and goods our economy would not suffer from the sudden drop in defense spending.

With approximately forty billion dollars yearly assistance to the have-not nations, their standard would be raised sufficiently to provide a much larger market for business and commerce for the advanced nations.

Failure of such a plan for a gigantic help will inevitably drive such nations to chaos, and to totalitarian system, either communist or fascist.

Peoples of the world, even the so-called backward peoples are aware now that a better life exists than their own, and feel that they are entitled to a better life.

They know that industrialization is the answer, that capital has to be gotten, or created, and are impatiently searching methods to speed up their advancement.

Where is the tremendous amount to come from?

Private capital is justly unwilling to invade most of those countries, the fate of Dutch investments in Indonesia is a sharp reminder, and there are others in various other countries.

All that remains for them is to lift themselves by their own bootstraps, to take out part of the daily bread from the mouth of their own peoples, even as Russia had to do, and is still forced to do, to carry the armament burden.

Such sacrifice could only be obtained by force, for the daily bread is all too small in those countries.

Do we prefer them to follow this road?

The road that inevitably leads them to the Soviet camp?

Yes, it is time for sober evaluation of the world, and it is time for greatness.

Newspaper clipping showing italicized passages in Himler's editorial. *Hungarian Miners' Journal*, January 2, 1958.

NOTES

Introduction

1. Ken Fones-Wolfe and Ronald L. Lewis discuss the transnational migration of communities to the Appalachian state of West Virginia, the establishment of ethnic communities, and the economic changes wrought on immigrants and West Virginia and by implication the entire Appalachian South in *Transnational West Virginia: Ethnic Communities and Economic Change, 1840–1940* (Morgantown: West Virginia UP, 2002). For additional information on Appalachia and its history, see Richard B. Drake, *A History of Appalachia* (Lexington: UP of Kentucky, 2001), which provides a general history of the southern Appalachian region. Other classic sources on Appalachia include David Whisnant, *All That is Native and Fine: The Politics of Culture in an American Region* (Chapel Hill: U of North Carolina P, 1983); Alan Batteau, ed., *Appalachia and America: Autonomy and Regional Dependence* (Lexington: UP of Kentucky, 1983); John Gaventa, *Power and Powerlessness: Quiescence and Rebellion in an Appalachian Valley* (Urbana: U of Illinois P, 1980); John R. Burch Jr., *Owsley County, Kentucky, and the Perpetuation of Poverty* (London: McFarland & Company, 2008) examines the roots of poverty in Eastern Kentucky. A few articles have specifically discussed the new immigration in particular areas of southern Appalachia. See, for example, Doug Cantrell, "Immigrants and Community in Harlan County, 1910–1930," *Register of the Kentucky Historical Society*, Vol. 88, No. 2, 119–41; Doug Cantrell, "Himlerville: Hungarian Cooperative Mining in Kentucky," *The Filson Club History Quarterly*, Vol. 66, No. 4, 513–42; Margaret Ripley Wolfe, "Aliens in Southern Appalachia, 1900–1920: The Italian Experience in Wise County, Virginia," *Virginia Magazine of History and Biography*, Vol. 87, 455–72; Margaret Ripley Wolfe, "Aliens in Appalachia: The Construction of the Clinchfield Railroad and the Italian Experience," in Emmett M. Essin III, ed., *Appalachia: Family Traditions in Transition* (Johnson City: East Tennessee State UP, 1975); Margaret Ripley Wolfe, "Aliens in Southern Appalachia: Catholics in Coal Camps," *Appalachian Heritage*, Vol. 6, 1978, 42–56; Margaret Ripley Wolfe, "Putting Them in Their Places: Industrial Housing in Southern Appalachia, 1900–1930," *Appalachian Heritage*, Vol 7, 1979, 27–36.

2. Figure on the population of immigrants in the Appalachian coal producing regions are found in the *Eleventh Census of the United States, Report on Population of the United States, 1890* (Washington: Govt. Printing Office, 1895); *Twelfth Census of the United States, Population, 1900* (Washington: Govt. Printing Office, 1901); *Thirteenth*

Census of the United States, Population, 1910 (Washington: Govt. Printing Office, 1911); *Fourteenth Census of the United States, Population 1920* (Washington: Govt. Printing Office, 1922); *Fifteenth Census of the United States, 1930* (Washington: Govt. Printing Office, 1932); U. S. Senate, *Reports of the Immigration Commission, Immigrants in Industry, The Bituminous Coal Industry*, (Washington: Govt. Printing Office, 1911).

3. The idea that southern Appalachia was geographically and culturally isolated from the rest of America has been discussed extensively in numerous books. Henry David Shapiro in *Appalachia on Our Mind: The Southern Mountains and Mountaineers in the American Consciousness, 1870–1920* (Chapel Hill: U of North Carolina P, 1978) examines the discovery of Appalachia by local color writers, scholars, industrialists, and others and shows how these people shaped the view that Americans had of Appalachia and Appalachians. Another good source that examines the view that Americans had of Appalachians is Allen W. Batteau, *The Invention of Appalachia* (Tucson: U of Arizona P, 1990); Richard A. Straw and H. Tyler Blethen, *High Mountain Rising: Appalachia in Time and Place* (Urbana: U of Illinois P, 2004) offers a look at Appalachia people across different time periods. Early sources that deal with the image of Appalachia include: Cratis Dearl Williams, "The Southern Mountaineer in Fact and Fiction" (Ph.D. Diss., New York Univ., 1961); Henry David Shapiro, "A Strange Land and Peculiar People: The Discovery of Appalachia, 1870–1920" (Ph.D. diss., Rutgers Univ., 1966); William Goddell Frost, "Our Contemporary Ancestors in the Southern Mountains," *Atlantic Monthly* 83 (March 1899); John C. Campbell, *The Southern Highlander and His Homeland* (Lexington: U of Kentucky P, 1969); Jack E. Weller, *Yesterday's People: Life in Contemporary Appalachia* (Lexington: U of Kentucky P, 1965); Horace Kephart, *Our Southern Highlanders: A Narrative of Adventure in the Southern Appalachians and a Study of Life Among the Mountaineers* (New York: Macmillan, 1913; reprint Knoxville: U of Tennessee P, 1976 and 1984); Bruce and Nancy Roberts, *Where Time Stood Still: A Portrait of Appalachia* (New York: Macmillan, 1970); Arnold Toynbee, *A Study of History* (New York: Oxford UP, 1947).

4. The ownership of Appalachia's mineral resources and life within the company towns throughout the region has been extensively documented and discussed in numerous sources, including Harry M. Caudill, *Night Comes to the Cumberlands: A Biography of a Depressed Area* (Boston: Little, Brown, and Company, 1962); Ronald D. Eller, *Miners, Millhands, and Mountaineers: Industrialization of the Appalachian South, 1880–1930* (Knoxville: U of Tennessee P, 1982); The Appalachian Land Ownership Task Force, *Who Owns Appalachia? Landownership and Its Impact* (Lexington: UP of Kentucky, 1983); Harry M. Caudill, *Theirs be the Power: The Moguls of Eastern Kentucky* (Urbana: U of Illinois P, 1983); Crandall A. Shifflett, *Coal Towns: Life, Work, and Culture in Company Towns of Southern Appalachia, 1880–1960* (Knoxville: U of Tennessee P, 1991); David Alan Corbin, *Life, Work, and Rebellion in the Coal Fields: The Southern West Virginia Miners, 1880–1922* (Urbana: U of Illinois P, 1991); Rodger Cunningham, *Apples on the Flood: The Southern Mountain Experience* (Knoxville:

U of Tennessee P, 1987) traces the negative image of Appalachian people to its historical experiences.

5. American immigration has been discussed in numerous sources. Maldwyn Allen Jones, *American Immigration* (Chicago: The Univ. of Chicago Press, 1960) is a good early general history of immigration as is Oscar Handlin, *Boston's Immigrants, 1790–1880* (Cambridge: Harvard UP, 1991). The new immigration is discussed in particular in Alan M. Kraut, *The Huddled Masses: The Immigrant in American Society, 1880–1921* (Arlington Heights, IL: Harlan Davidson, 1982); Humbert Nelli, *The Italians in Chicago, 1880–1930* (New York: Oxford UP, 1970); Thomas Kessner, *The Golden Door: Italian and Jewish Immigrant Mobility in New York City, 1880–1915* (New York: Oxford UP, 1977); Glen C. Altschler, *Race, Ethnicity, and Class in American Social Thought, 1865–1919* (Arlington Heights, IL: Harlan Davidson, 1982); Hunbert Nelli, *From Immigrants to Ethnics: The Italian Americans* (New York: Oxford UP, 1983); Jacob Riis, *How the Other Half Lives* (Boston: Bedford/St. Martin's, 1996) provides a class photo/text look at immigrant life in urban America; J. Campbell Bruce, *The Golden Door: The Irony of Our Immigration Policy* (New York: Random House, 1954) examines American immigration policy through the Second Red Scare.

6. An older look at why immigrants came to the United States is found in Eleanor B. Tripp, *To America: The Story of Why People Left their Homes for the New Land* (New York: Harcourt, Brace, and World, 1969); An examination of Catholicism among American immigrants is found in James T. Fisher, *Communion of Immigrants: A History of Catholics in America* (New York: Oxford UP, 2002); A look at Jewish immigration to the United States is found in Ruth Gay, *Unfinished People: Eastern European Jews* (New York: W. W. Norton & Company, 1996); Deborah R. Weiner, *Coalfield Jews: An Appalachian History* (Urbana: U of Illinois P, 2006) provides a look at Jews in the Appalachian coalfields; Deno Trakas, *Because Memory Isn't Eternal: The Story of Greeks in Upstate South Carolina* (Spartanburg: Hub City Press, 2010) recounts the migration of his Greek family to the American South.

7. Good sources on nativism and prejudice against immigrants include Reed Ueda, *PostWar Immigrant America: A Social History* (Boston: Bedford/St. Martins, 1994); Leonard Dinnerstein, Roger Nichols, and David Reimers, *Natives and Strangers: Ethnic Groups and the Building of America* (New York: Oxford UP, 1979).

8. John Bodnar, *Workers' World: Kinship, Community, and Protest in an Industrial Society, 1900–1940* (Baltimore: Johns Hopkins UP, 1982) examines the immigrant work experience in the United States; Arthur C. Prichard, "Two Hundred Pounds or More: The Lebanese Community in Mannington," *Goldenseal*, Vol. 4, 2–3 (Sept. 1978), 18–24, recounts the experiences of pack peddlers in Appalachian mining communities.

9. See Oscar Handlin, *The Uprooted* (Boston: Little, Brown & Company, 1951) for a classic interpretation that immigrants came to America after being uprooted in their native lands; Nelli, *From Immigrant to Ethnic* is a good look at how immigrants created

and used ethnic institutions to ease the transition to industrial life; the first chapter of Herbert Gutman, *Work Culture and Society in Industrializing America* (New York: Vintage Books, 1976) discusses how immigrant institutions developed to ease the transition from an agricultural society to an industrial one.

10. Victor R. Greene, *American Immigrant Leaders: Marginality and Identity, 1800–1910* (Baltimore: Johns Hopkins UP, 1987) challenges the melting pot theories and the idea of studying immigrant groups from the bottom and contends that immigrant leaders helped foreigners assimilate into American culture through the promotion of ideas of Americanism within the ethnic community; John Higham, ed., *Ethnic Leadership in America* (Baltimore: Johns Hopkins UP, 1978) examines the different styles and approaches to immigrant leadership among ethnic groups in the United States.

11. Ellen Schrecker, *The Age of McCarthyism: A Brief History with Documents* (Boston: Bedford/St. Martins, 1994) offers a good examination of the Second Red Scare that followed World War II.

12. Himler details these fights within various issues of *Magyar Bányászlap*, the Hungarian language newspaper he published from 1910 until the 1950s. He also makes reference to the incidents in the autobiography.

13. A number of sources document the Progressive Era and the New Deal. Classics include Eric Goldman, *Rendezvous with Destiny* (New York: Knopf, 1953); Richard Hofstadter, *The Age of Reform* (New York: Vintage Books, 1955); Arthur Link, *Woodrow Wilson and the Progressive Era, 1910–1917* (New York: Harper & Row, 1954); George Mowry, *The Era of Theodore Roosevelt and the Birth of Modern America, 1900–1912* (New York: Harper & Row, 1958; William Leuchtenberg, *The Perils of Prosperity, 1914–1932* (Chicago: U of Chicago P, 1958).

1. Mátraverebély

1. Pásztó is a small town with around 10,000 inhabitants at the western foot of the Mátra Mountains in Nógrád County, Hungary. *Irány Magyarország!*, www.1hungary.com.

2. Mátraverebély is a village with around 2,000 inhabitants at the northwestern foot of the Mátra Hills in Nógrád County, Hungary. *Irány Magyarország!*, www.1hungary.com.

3. An Orthodox Jew follows a modern-day branch of the Jewish religion in which Orthodox Judaism is the most traditional expression of modern Judaism when compared to Reform Judaism and Conservative Judaism. Orthodox Jews believe the entire Torah- the "written law" (i.e. the five books of Moses) given by God at Sinai- remains authoritative for modern life in its entirety. Orthodox Jews also hold to the "oral law" (i.e. the Talmud). *ReligionFacts*, www.religionfacts.com.

4. Yiddish is the historic language of Ashkenazic Jews of Central and Eastern Europe. Yiddish results from a fusion of elements derived principally from medieval German dialects and secondarily from Hebrew and Aramaic, various Slavic languages,

and Old French and Old Italian. *The American Heritage Dictionary of the English Language*, 5th ed (Boston: Houghton Mifflin Harcourt, 2011), 2010.

5. Hail Mary is a prayer for the intercession of the Blessed Virgin Mary, mother of Jesus Christ. The Hail Mary is based on these passages from the Bible: Luke 1:28–35; 42–48. *Catholic Planet*, www.catholicplanet.com.

6. A Gentile is a person who is not Jewish. *The American Heritage Dictionary of the English Language*, 5th ed (Boston: Houghton Mifflin Harcourt, 2011), 733.

7. The Torah is the whole body of the Jewish sacred writings and tradition, including the oral tradition. AudioEnglish.org.

8. The structure of education in nineteenth and twentieth century Hungary and in Eastern Europe was different than our modern educational structure. Hungarian students in the late nineteenth and early twentieth centuries attended a compulsory elementary school for grades one through four, then moved ahead to a compulsory high school which encompassed separate grades one through five. The former Hungarian high school grades one through five equal grades five through nine in today's education structure of both Hungary and the United States. When Hungarian students finished grade nine, they completed their high school education by attending the Academy of Commerce, primarily a trade/vocational school, or Gymnasium or Real School. Gymnasium and Real School were senior secondary schools that prepared students for attending a university and that included Hungary's and America's current tenth through twelfth grades. Willem Frijhoff, "Gymnasium Schooling" in *Encyclopedia of Children and Childhood in History and Society*, University of Cincinnati (New York: Macmillan Reference, USA, 2004), 307, www.researchgate.net, and Thom Hartmann, "The 'Real' School Is Not Free," *Thom Hartmann Program*, November 1, 2007, www.thomhartmann.com.

9. Turócszentmárton was a town in Slovakia, now called Svätý Martin, and was part of the Kingdom of Hungary from the second half of the tenth century until the 1920 Treaty of Trianon. Svätý Martin has a population of approximately 61,000 and is the eighth largest city in Slovakia. Mongabay, Mongabay.com, and www.hunmagyar.org.

10. Cognac is a brandy distilled from white wine and produced in the vicinity of Cognac. *The American Heritage Dictionary of the English Language*, 5th ed (Boston: Houghton Mifflin Harcourt, 2011), 358.

11. Losonc was a town in Slovakia, now called Lucenec, that was the eighteenth century capital of Nógrád County of the Kingdom of Hungary. Lucenec became part of Czechoslovakia as a result of the 1920 Treaty of Trianon and part of Slovakia in the 1993 peaceful dissolution of Czechoslovakia into the Czech Republic and Slovakia. Travel in Slovakia, www.slovakia.travel, and www.hunmagyar.org.

2. America

1. Tsarist Russian was an inherited autocratic form of government that ended with the death of Tsar Nicholas II in the 1917 Bolshevik Revolution. *National Geographic* online, www.nationalgeographic.com.

2. The term "Golden Age" refers to a period of great peace, prosperity, and happiness. In Greek and Roman mythology, "Golden Age" is the first age of the world, an untroubled and prosperous era during which people lived in ideal happiness. *The American Heritage Dictionary of the English Language*, 5th ed (Boston: Houghton Mifflin Harcourt, 2011), 755.

3. "Proletars" is the Latin term used to refer to Roman citizens of the lowest class who contribute to the state only through having children. *The American Heritage Dictionary of the English Language*, 5th ed (Boston: Houghton Mifflin Harcourt, 2011), 1409.

4. The *Carpathia* was built in 1902 by C.S. Swan and Hunter, with Wallsend-on-Tyne (engines by Wallsend Slipway Co. Ltd) for the Cunard SS Company. The *Carpathia* was a 13,555 gross ton ship, length 540ft. x beam 64.5ft., one funnel, four masts, twin screw, speed of 14 knots. Passenger accommodations were for 204-2nd and 1,500-3rd class passengers. The *Carpathia* was refitted in 1905 to carry 100-1st, 200-2nd, and 2,250-3rd class passengers and then resumed sailings between Trieste, Fiume, Palermo, and New York on Oct. 27, 1905. The *Carpathia's* first voyage was from Liverpool to Queenstown (Cobh) and Boston on May 28, 1903, and the last voyage was on July 17, 1918, when the ship was torpedoed and sunk by the German submarine U.55, 120 miles west of Fastnet. Five lives were lost in this torpedo attack. The *Carpathia* is best known for its role as the rescue ship for passengers of the *Titanic* when the *Titanic* sank off Newfoundland on its first voyage from Britain to New York in April, 1912. The *Carpathia* arrived in New York on April 18, 2012, with 700 *Titanic* survivors. NRP Bonsor, *North Atlantic Seaway*. vol.1, p.156, on *TheShipsList*, CARPATHIA 1902, www.theshipslist.com, and "*Carpathia* (R.M.S.), Wreck of the *Carpathia*, *Titanic's* Rescuer, Found." *Reuters Limited*, Copyright 2000, Sept.22. Article located on National Underwater and Marine Agency website, www.numa.net.

5. Abbázia is the old Hungarian name for Opatija, the leading tourist destination in the Republic of Croatia. Opatija is the first climatic seaside resort on the Adriatic, and its beaches have attracted visitors for more than 160 years. "The 'must see' sights of Opatija." Visit Opatija, www.visitopatija.net.

6. The Purser's Office is the location on a ship where matters relating to money and comfort for the passengers and crew are handled. *Merriam-Webster* online, www.merriam-webster.com.

7. "As heady as champagne," Austrian-Hungarian composer Franz Lehár's 1905 classic and effervescent operetta, *The Merry Widow*, ("Die Lustige Witwe"), achieved world-wide success. The libretto was written by Viktor Léon and Leo Stein, and Lehár created a new style of Viennese operetta by introducing waltz tunes, imitations of Parisian cancan dances, and a certain satirical element. Boston Lyric Opera website, blo.org.

8. Naples is a city of south-central Italy on the Bay of Naples, an arm of the Tyrrhenian Sea. Naples is a major seaport and a commercial, cultural, and tourist center.

The American Heritage Dictionary of the English Language, 5th ed (Boston: Houghton Mifflin Harcourt, 2011), 1171.

9. Palermo is a city of north-west Sicily, Italy, on the Tyrrhenian Sea. *The American Heritage Dictionary of the English Language*, 5th ed (Boston: Houghton Mifflin Harcourt, 2011), 1269.

10. Gibraltar is a British colony centered around the heavily fortified Rock of Gibraltar, a strategically located peninsula on the north side of the Strait of Gibraltar, connecting the Mediterranean Sea and the Atlantic ocean between Spain and northern Africa. *The American Heritage Dictionary of the English Language*, 5th ed (Boston: Houghton Mifflin Harcourt, 2011), 741.

11. Ellis Island, New Jersey, opened in 1892 and served as a federal immigration station until the station closed in 1954. From 1900 to 1914, the peak years of Ellis Island's operation, some 5,000–10,000 people passed through the immigration station every day. Approximately 80 percent successfully passed through in a matter of hours, but others could be detained for days or weeks. HISTORY, www.history.com.

12. Trachoma is a bacterial infection that affects the eyes. Trachoma is contagious and spreads through contact with the eyes, eyelids, and nose or throat secretions of infected people. Trachoma is the leading preventable cause of blindness worldwide. Mayo Clinic. www.mayoclinic.com.

13. "How much money do you have?" Reverso Dictionary, dictionary.reverso.net/german-english/.

3. Investment in America

1. A schooner is a large beer glass, generally holding a pint or more. *The American Heritage Dictionary of the English Language*, 5th ed (Boston: Houghton Mifflin Harcourt, 2011), 1570.

2. "Thacker Mines Topo Map in Mingo County WVA." Thacker Mines is a small, unincorporated place name approximately 2.5 miles NE of Thacker, WV. Thacker Mines is situated on Thacker Creek, a tributary to the Tug Fork of the Big Sandy River in Mingo County, WV. USGS Topo Map Quad: Majestic. *TOPOZONE*, www.topozone.com.

3. "Witty, poetic, erotic, and brutal, *One Thousand and One Nights* are the never-ending stories told by the young woman Shahrazad under sentence of death to King Shahrayar. Maddened by the discovery of his former wives' unfaithfulness, King Shahrayar believes all women are unfaithful and vows to marry a virgin every night and kill her the following morning. To survive, King Shahrayar's newest wife, Shahrazad, spins a web of mesmerizing tales night after night, leaving the ending of each tale till the following evening, and thus prolonging her life for another day." The tales of *One Thousand and One Nights* were originally written in Arabic and were gathered from tales told in Persia, India, and the Great Arab Empire. Al-Shaykh, Hanan, comp., and

Mary Gaitskill (Introduction), *One Thousand and One Nights: A Retelling* (New York: Pantheon, June 11, 2013).

4. John Freddie Wilson, "Coal Mine Scrip." In the early 1900's, many of the coal mines were located in remote areas. Coal companies and mine owners capitalized on this opportunity of isolation to make additional profits by building company stores and creating the monetary system called "scrip." The scrip system forced the miners to be able to use only scrip, rather than American currency, at the company stores. Scrip manufacturers minted scrip in brass, copper, aluminum, zinc, and also produced rubber scrip of various denominations. Hosted by *rootsweb* on ancestry.com community, 2006. Page updated by Lynda Combs Gipson, 2009. www.rootsweb.ancestry.com.

5. A drift mouth is the opening or entry to a drift mine, an underground coal mine that is accessed on the slope of a hill and that follows the underlying coal seam in a horizontal direction. Coal miners may call the opening of a coal mine a drift mouth, even though all coal mines aren't structured as a drift mine. Tom W. Corbin (geologist) in discussion with editor, December 2015.

6. Glen Alum is a small place name approximately 5 miles SE of Thacker Mines, Mingo County, West Virginia. *Topo Quest,* USGS Map Name: Whamcliffe, WV, topoquest.com.

7. Portsmouth, Ohio, located at the mouth of the Scioto River and the Ohio River, was the most important commercial and manufacturing city on the Ohio River between Wheeling, West Virginia, and Cincinnati, Ohio. In 1916, Portsmouth was listed as being a major industrial and jobbing center. There were nearly one hundred manufacturing enterprises located in the city of Portsmouth, including the fourth-largest shoe manufacturer in the United States, which turned out over twenty thousand shoes a day from a half dozen modern plants. Portsmouth manufactured more paving and fire brick than any city in the United States, and Portsmouth was home to a large steel plant employing several thousand men. Portsmouth's numerous furniture and wood-working factories, planing mills, and machine shops employed over ten thousand men. The Norfolk and Western and three other lines of railway supplied transportation infrastructure for the Portsmouth manufacturing industries. *Norfolk and Western Railway Industrial and Shippers Guide. Compiled by the* Agriculture and Industrial Dept. N.& W. Ry (Roanoke: Union Printing and Manufacturing Company, 1916).

8. The Russo-Japanese Treaty is the agreement signed by Russia and Japan and mediated by U.S. President Theodore Roosevelt at Portsmouth, New Hampshire, in August, 1905. The treaty brought the end to the Russo-Japanese war fought by Russia and Japan as Russia resisted Japan's imperial designs for East Asia. "Russo-Japanese Peace Treaty Signed," HISTORY, www.history.com.

4. I Became an American

1. The imposing Buda King's Castle on Castle Hill (Buda Mountain) overlooks Budapest from its elevated position on Buda Mountain, rising 48 meters above the

Danube River. The foundations of today's King's Castle, which was besieged no less than 31 times, were laid in the fourteenth century by King Lajos the Great. *Aviewoncities*, www.aviewoncities.com.

2. *Alice in Wonderland*, first known as *Alice's Adventures in Wonderland*, was written by Charles Ludwidge Dodgson, best known by his pseudonym, Lewis Carroll. Carroll was born in Daresbury, England, on Jan. 27, 1832, and Carroll was quite adept at creating imaginative stories to entertain his 10 siblings and himself. As an adult, Carroll loved to entertain children, and he told the first iteration of *Alice's Adventures in Wonderland* to a picnic for the young Alice Liddell and her sisters. By 1932, *Alice's Adventures in Wonderland* was one of the most popular books in the world. "Lewis Carroll Biography," BIOGRAPHY, www.biography.com.

3. The St. Louis Exposition also is known as the 1904 World's Fair. The magnificent fairground equated America's expansion westward since the Louisiana Purchase with the nation's cultural and economic progress. By the time that the Fair closed on December 1, 1904, an estimated 20 million people had visited the Fair. Missouri Historical Society, www.mohistory.org.

4. The Ku Klux Klan was a secret society organized in the South after the Civil War to reassert white supremacy by means of terrorism. *The American Heritage Dictionary of the English Language*, 5th ed (Boston: Houghton Mifflin Harcourt, 2011), 978.

5. The Silver Shirters were a pro-Nazi group led by notorious American Fascist, William Dudley Pelley. As leader of the Silver Shirters, Pelley preached a toxic brew of anti-semitism, nationalism, and mysticism. www.encyclopediaofarkansas, and Jack Bradstreet, IMDb Mini Biography of William Dudley Pelley, www.imdb.com.

6. The Black Legion was founded in the mid- 1920's as the Black Guards, a security force for officers of the Ohio Ku Klux Klan. The Legion's political objectives were broad and, at the same time, narrowly specific. As one of the Legion's promotional pieces stated, "We will fight political Romanism [The Catholic Church], Judaism, Communism, and all 'isms' which our forefathers came to this country to avoid." "History: The Black Legion, Where Vets and The Klan Met," Veterans Today, November 28, 2012, www.veteranstoday.com.

7. Gingham is a printed or dyed fabric that is known for its checked patterns of white and a bold color. Originally, this fabric was intended to be woven into stripes. Its name comes from the word "genggang," which means "stripes" in the Malay language spoken in places such as Malaysia and Indonesia. WiseGeek/Clear Answers for Common Questions, www.wisegeek.org.

8. Pig iron is a semi-finished metal produced from iron ore in a blast furnace. Pig iron contains 92 percent iron, a high amount of carbon (typically up to 3.5 percent), and a balance of largely manganese and silicon, plus small amounts of phosphorous, sulfur, and other impurities. Pig iron is further refined in a furnace for conversion into steel. Pig iron gets its name from the shape of the trough (resembling a pig) in which the iron once was cast in the nineteenth century. www.businessdictionary.com.

9. Slag is the waste content of metal ores that floats on the surface during the process of extracting or refining metal. Containing mainly silicates, sulfates, and phosphates of calcium, slag is used principally in road building and concrete admixtures. www.businessdictionary.com.

5. Cleveland, Ohio

1. The Park and Tilford grocery stores were upscale businesses founded by Joseph Park (1823–1903) and John Mason Tilford (1815–1891). The first Park and Tilford store was located at 35 Carmine Street, Manhattan, and by 1909, the New York telephone directory listed eight Park and Tilford locations. The 310 Lenox Avenue location of Park and Tilford is built in the Classical Revival style and was constructed in 1908. Architects for this construction were Simonson and B. Hustan. The 310 Lenox Avenue store is listed on the National Register of Historic Places and, as of 2014, this Park and Tilford store was home to the Red Rooster Restaurant and other shops. waltergrutchfield.net, National Park Service/National Register of Historic Places, focus.nps.gov.

2. Goulash is a stew of beef or veal and vegetables seasoned mainly with paprika. *The American Heritage Dictionary of the English Language*, 5th ed (Boston: Houghton Mifflin Harcourt, 2011), 760.

3. "Hunkies" is an alteration of Hungarian. The first known use of "Hunky" was circa 1896 and usually was used as a disparaging term for persons of central or east European birth or descent. *Merriam-Webster Online*, www.merriamwebster.com.

4. "Hunyaks" is an early twentieth century North American informal, offensive alteration of the name of persons of Hungarian or central European origin, especially immigrants. *Oxford Dictionaries Online*, www.oxforddictionaries.com.

5. "Dagoes" is an offensive slang used as a disparaging term for a person of Italian, Spanish, or Portuguese descent. *The American Heritage Dictionary of the English Language*, 5th ed (Boston: Houghton Mifflin Harcourt, 2011), 456.

6. The Daughters of the American Revolution (DAR) was founded in 1890 with the simple mission of promoting historic preservation, education, and patriotism. DAR is a nonprofit, nonpolitical, volunteer women's service organization consisting of over 180,000 members in 3,000 chapters across the world. Daughters of the American Revolution, www.dar.org.

6. Again in Coal Mine

1. The Tangier American Legation, located in Tangier, Morocco, was the first American-owned government property outside the boarders of the United States. The property was gifted to the American government by Sultan Moulay Suleiman in 1821, and the property played a key role in the success of the 1942 Allies Operation Torch landings in Morocco and Algeria. The site also was pivotal in the evacuation

of thousands of Jewish refugees from occupied Europe throughout the war. David Weible, "American History on Foreign Soil: The Tangier American Legation in Morocco." National Trust for Historic Preservation, May 28, 2014, savingplaces.org.

2. J.P. Morgan was one of the world's greatest industrial and financial leaders of the twentieth century. J.P. Morgan built the United States Steel Corporation by combining ten different steel companies, including the two largest, Carnegie Steel and the Federal Steel Company, into a trust. A trust is formed when different companies in an industry combine to reduce competition and increase profits. "United States Steel Corporation," Reference For Business, www.referenceforbusiness.com.

3. The 1892 Homestead Strike in Pennsylvania's Carnegie Steel Co. pitted union labor wages and work rules against Carnegie's desire to cut costs. Carnegie Steel hired the Pinkerton National Detective Agency to help break the union. At the end of the battle, seven workers and three Pinkerton officers were dead. Four days later, 8,500 National Guard forces were sent to control order. "1892 Homestead Strike," AFL-CIO, America's Unions, www.aflcio.org.

4. Ribs are the sides of a pillar (an area of coal left to support the overlying strata in a mine) sometimes left permanently to support surface structures or to support the wall of an entry to the solid coal on the side of any underground coal mine passage. "Kentucky Coal and Energy Education Project," Glossary of Mining Terms, *Kentucky Coal Education*, www.coaleducation.org.

5. The precursor to America's Central Intelligence Agency, the Office of Strategic Services, was formed in June, 1942, by Presidential Directive from America's President Franklin D. Roosevelt. The OSS replaced the former American intelligence system, Office of the Coordinator of Information (OCI), that was considered to be ineffective. Roosevelt selected Colonel William Donovan as the first director of the OSS, and Donovan was given the rank of Major General. During WW II, Donovan built up a team of 16,000 agents working behind enemy lines. The OSS was disbanded in October, 1945, and eventually was replaced by the Central Intelligence Agency (CIA). "Office of Strategic Services," *Spartacus Educational,* spartacus-educational.com., Bruce L. Brager, "The Office of Strategic Services," Military History Online, On a March Through the Past, www.militaryhistoryonline.com., and Duncan Bare, "Hungarian Affairs of the US-Office of Strategic Services in the Mediterranean Theatre of Operations from June 1944 until September 1945," Section 3.1, p.33, MA Thesis, University of Graz. Thesis on file with editor.

6. The Bowery is a section of lower Manhattan in New York City. The Bowery once was notorious for its saloons, petty criminals, and derelicts. *The American Heritage Dictionary of the English Language*, 5th ed (Boston: Houghton Mifflin Harcourt, 2011), 219.

7. The Civil Works Administration was authorized under Franklin D. Roosevelt's New Deal in 1933. The CWA was a jobs program for the unemployed, provided employment for 2.5 million in a month's time, and eventually reached an employment total of 4 million. "Putting People Back to Work," U.S. History, www.ushistory.org.

7. Riding the Rods

1. Mulligan is a stew of various meats and vegetables and is also called "mulligan stew". *The American Heritage Dictionary of the English Language*, 5th ed (Boston: Houghton Mifflin Harcourt, 2011), 1156.

2. A doyen is the eldest or senior member of a group. *The American Heritage Dictionary of the English Language*, 5th ed (Boston: Houghton Mifflin Harcourt, 2011), 542.

3. Sir William Blackstone, (1723–1780), is the great eighteenth century English legal scholar whose philosophy and writings were infused with Judeo-Christian principles. The Ten Commandments are at the heart of Blackstone's philosophy. Blackstone taught that man is created by God, granted fundamental rights by God, and man's law must be based on God's law. Blackstone's great work, *Commentaries on the Laws of England*, was basic to the U.S. Constitution and was the basic textbook of America's early lawyers. It was only in the mid-twentieth century that American law, being re-written by the U.S. Supreme Court, repudiated Blackstone. An attack on Blackstone is an attack on the U.S. Constitution and America's Judeo-Christian foundations. The Blackstone Institute is committed to reviving the Constitution and its Blackstonian foundations. "Sir William Blackstone," Blackstone Institute, www.blackstoneinstitute.org.

4. Chinese Kanakas is the U.S. nautical and Australian name for native South Sea islanders. The name is taken from the *Hawaiian* "kanaka" which means "man." "Word origin and History for kanaka," *Dictionary.com*, www.dictionary.reference.com.

5. Sarongs are garments consisting of a length of cloth wrapped around the body and tied at the waist or below the armpits, and are worn by men and women in Malaysia, Indonesia, and the Pacific islands. *The American Heritage Dictionary of the English Language*, 5th ed (Boston: Houghton Mifflin Harcourt, 2011), 1558.

8. Romance

1. Calumet, Michigan, is located in Michigan's Upper Peninsula. In 1890, Calumet was in the heart of the Copper Country, and 60,000 people were employed in Calumet. Now the headquarters of the Keweenaw Historical Park, Calumet is working to not only preserve its rich history and architecture, but also to restore the vibrancy of the community through its Main Street Program. Pure Michigan, www.michigan.org.

2. "Swabian" refers to a reference from a region and medieval duchy in Southwest Germany that constituted the area presently included in the states of Baden-Württemberg and Bavaria in South Germany. Dictionary.com, www.dictionary.reference.com.

3. Calcimine is a whitewash or whitewash with zinc oxide or other pigments added, formerly used as a coating for plaster walls and ceilings. *The American Heritage Dictionary of the English Language*, 5th ed (Boston: Houghton Mifflin Harcourt, 2011), 262.

4. The McAdoo Tunnel construction under the Hudson River connected New Jersey with New York City, and the tunnel construction was begun by the Hudson Companies in 1902. The tunnel was built by pushing a mechanical shield through the

silt at the bottom of the river. The displaced mud then was carried to the surface in small cars. All work was done under an air pressure of thirty-eight pounds per square inch. The Hudson Companies spent the first two months of 1908 testing the trains on the Second Avenue Elevated and running the trains, loaded with enough sandbags to equal the weight of a full load of passengers, through the tunnel tubes. "The Public Be Pleased: William Gibbs McAdoo and the Hudson Tubes," www.nycsubway.org.

9. In the Show Business

1. The Baldwin-Felts Agency was formed in 1900 as a protective agency for the Norfolk and Western and other railroads in Southwest Virginia, Eastern Kentucky, and West Virginia. Later, the agency became involved in labor/union struggles in the coalfields by enforcing law and order for the mine operators in areas where state police and county sheriffs were unable to police the many mine camps. Baldwin-Felts dissolved in 1930. "Guns, Thugs, and Heroes," The Roanoker, July, 1979, theroanoker.com.

11. On the Road Again

1. Meerschaum is a fine, compact, usually white claylike mineral of hydrous magnesium silicate, $H_4Mg_2Si_3O_{10}$, found in the Mediterranean area, and used in fashioning tobacco pipes and as a building stone. *The American Heritage Dictionary of the English Language*, 5th ed (Boston: Houghton Mifflin Harcourt, 2011), 1094.

2. The Hocking Coalfield of Ohio was one of the most important coalfields in Northern Appalachia during the coal mining boom from approximately 1880 to 1920. The main coal seam mined in the Hocking Coalfield was the Middle Kittanning (called No. 6 coal in Ohio) which was at least 60" thick. The Hocking Coal Field included the following Ohio Counties: Perry, Hocking, and Athens. Coalfields of the Appalachian Mountains, www.coalcampusa.com.

3. *Kismet* is most famous as a musical but didn't actually become a musical until 1953 when Robert Wright and William (Wilhelm) Dieterle adapted the *Kismet* story four decades after *Kismet*'s debut. The original *Kismet* was a play that opened in London in 1911 and was adapted for the silent film screen in 1914. Additional non-musical remakes of *Kismet* were produced in 1920, 1930, and 1944, with the 1944 re-make done in vivid color. The second film version of *Kismet* stars Otis Skinner, the leading man of *Kismet*'s first New York production, as Hajj, a rascal beggar who asks for alms outside a mosque in Bagdad. Hajj eventually schemes to marry his daughter to royalty and to win the heart of the queen of the castle himself. Fritzi Kramer, Movies Silently, March 9, 2014, moviessilently.com, and Letterboxd, *Kismet* 1920, Directed by Louis J. Gasnie, letterboxd.com.

4. The Dayton Flood of 1913 began on Easter Sunday, March 23, 1913. "Torrential rains across the Midwest dropped a record equivalent three months of rainfall within four days. Floodwaters funneled down Ohio's Miami Valley into the heart of the

vibrant industrial city of Dayton. Levees burst, houses were swept away, and downtown Dayton was gutted by fires blazing from broken gas mains. At the end of Easter week, nearly 100 Daytonians had perished, and tens of thousands more were left homeless and destitute- a tragedy that made banner headline in newspapers nationwide." Trudy E. Bell, "The Great Dayton Flood of 1913," *Images of America* (Ohio: Arcadia Publishing, 2008).

5. The Y Bridge of Zanesville is located at the confluence of the Licking and Muskingham Rivers. Downtown Zanesville had been plagued by flooding problems since 1814, and engineers determined that the long- sought solution to Zanesville's watery woes was the construction of a unique Y-shaped bridge. So remarkable is the bridge design that pioneering aviatrix Amelia Earhart remarked that Zanesville is "the most recognizable city in the country." Zanesville's charming strange bridge stands among just a few bridges in the world that can be crossed without changing sides of the river. "Zanesville Y-Bridge," Atlas Obscura, www.atlasobscura.com.

6. The poem "Worthwhile" is one of the *Poems of Sentiment* composed by Ella Wheeler Wilcox and copyrighted in Chicago, IL, by W.B. Conkey Company, 1906. Wilcox, (Nov.5, 1850–Oct.30, 1919), was a prolific American poet, journalist, and free thinker. www.ellawheeler.org.

7. The *Ryndam*, a Holland American Line ship, was a 12,340 gross ton ship with length 550.3ft x beam 62.3ft, one funnel, two masts, twin screw, and a speed of 15 knots. The *Ryndam* accommodated 286-1st, 196-2nd, and 1,800-3rd class passengers and was launched on its first voyage on October 10, 1901, when it left Rotterdam for New York. The *Ryndam's* last Rotterdam-New York voyage was in 1929. NRP Bonsor, *North Atlantic Seaway*, vol.3, p.913, on The Ships List, "Ryndam 1901," www.theshipslist.com.

8. The Balkan Wars began on October 8, 1912, when the tiny Kingdom of Montenegro declared war on the weak Ottoman Empire, then under Turkish rule. Three other Balkan states—Bulgaria, Greece, and Serbia—joined the war on the Ottoman Empire and effectively pushed the enfeebled Ottomans out of Europe by March, 1913. By July, 1913, Greece and Serbia were clashing with Bulgaria in the Second Balkan War, a bitter month-long struggle that prompted Europe's great powers to become entwined in the fate of the Balkans. The Balkan struggles climaxed with the murder of the crown prince of the Austro-Hungarian Empire by a Yugoslav nationalist, and Europe plunged into WW I. Ishaan Tharoor, "The Balkan Wars: 100 Years Later, A History of Violence." *Time*, Oct. 8, 2012.

12. With Mother

1. Gendarmes are members of a French national police organization constituting a branch of the armed forces with responsibility for general law enforcement. *The American Heritage Dictionary of the English Language*, 5th ed (Boston: Houghton Mifflin Harcourt, 2011), 730.

2. Szombathely, founded in 45 AD, is the oldest city in Hungary. Following the Treaty of Trianon, Hungary lost many western territories to Austria, and Szombathely, being only ten kilometers from the new state border of Hungary, ceased to be the center of Western Hungary. *AVEC*, www.avecnet.net.

3. Debrecen is the second largest city of Hungary and is situated in the eastern part of Hungary. Debrecen is one of Hungary's most vivid cultural and social centers. Debrecen.hu, www.eng.debrecen.hu.

4. The *Pretoria* was built in 1897 for the Hamburg American Line. It was a 12,800 gross ton ship with a length of 560ft x beam 62ft, straight stem, one funnel, four masts, twin screw, and a speed of 14 knots. The *Pretoria* accommodated 162-1st, 197-2nd, and 2,382-3rd class passengers. In 1908, it collided with and sank the *NIPPONIA* in thick fog off Texel. The *Pretoria* was rebuilt in 1910 to 13,234 tons with passenger accommodations for 400-2nd and 2,200-3rd class passengers. PRN Bonsor, *North Atlantic Seaway*, vol.1, p.156, on The Ships List, "Pretoria 1897," www.theshipslist.com.

5. From 1881 until 1914, the Hamburg-American Line, based in Hamburg, Germany, was the largest shipping line in existence. It transported hundreds of thousands of emigrants from Germany, Scandinavia, and Eastern Europe to the United States and other destinations around the world. By 1914, when World War I erupted, the Hamburg-American Line had 175 active steamships, several of which could carry one thousand passengers each. Karen Manners Smith, "Hamburg-Amerika," immigration tounitedstates.org.

6. Karoly Fulop, (1893–1963), is recognized as one of the most prolific Hungarian-American artists and produced works in the mediums of paint, watercolor, ceramic, wood, and bronze. Fulop studied art in Budapest, Munich, and Paris. He was an artist of unique and curious influences who incorporated icons from his Catholic background with symbolist images from the Modern avant garde art movements. Fulop moved to New York in 1920 and then to Los Angeles in the 1930's. He opened a school of decorative arts in his Los Angeles studio and always maintained a studio in Paris. Fulop gained fame in the U.S. when he designed the murals for the Philadelphia Public Library. Several of Fulop's works are visible and available for sale on www.artnet.com. Papillion, www.papillongallery.com.

7. Ilona Fulop was one of Hungary's most acclaimed female journalists and screen writers. In a 1919 Fulop article entitled "What is 'Revolution' Doing to Love?", Miss Fulop describes herself as a socialist radical as opposed to a revolutionary or a Bolshevik. Her desire to further equal rights for women strongly defines her as a modern-day feminist. Royal Books, www.royalbooks.com.

8. *Spring Shower* is a film that was released as *Tavaszi zápor* in Hungary in 1932, as *Marie, Legende Hongroise* in France in 1933, and as *Spring Shower* in the U.S. in 1935. Ilona Fulop wrote the *Spring Shower* screenplay, and Hungarian film director, Pál Fejös, directed this screenplay that now is regarded as one of the all-time great films coming from Hungary. *Spring Shower* echoes Fulop's feminist views with its

women and proletarian—friendly plot centered around the fate of a poor Hungarian girl who is driven out of her village when she becomes pregnant by her employer's wealthy fiancé. A presentation copy of an early *Spring Shower* draft script, housed in a likely hand-sewn floral cloth-cover portfolio, is currently available for sale for $2,850.00 at www.royalbooks.com. This script copy is probably the only surviving copy of the script and includes a note from Fulop on the title page. The handwritten note states: "To Mac: Because you still trust me! Ilona, Christmas, 1929, Hollywood, Cal." Royal Books, www.royalbooks.com.

9. William Ashley "Billy" Sunday, (Nov. 19, 1862–Nov. 6, 1935), was an American evangelist known for his arresting 'fire and brimstone' preaching style. Billy left high school before he graduated. He played professional baseball for the Chicago White Stockings, and Billy boasted a career record 92 stolen bases which was topped only by Ty Cobb. In 1894, Sunday quit baseball, and in 1903, he became an ordained preacher. Sunday was credited with being a major social influence in the temperance movement which led to Prohibition in 1919. One of his most famous sermons was "Booze, or, Get on the Water Wagon," which persuaded many to give up drinking. Billy died a wealthy man. Most of his money was given to him by contributions during his charismatic sermonizing. United States History, www.u-s-history.com.

13. I Become a Publisher

1. The United Coal Miners' Union is today known as The United Mine Workers of America. The United Mine Workers of America was founded in 1890, and this union has a diverse membership that includes coal miners, clean coal technicians, health care workers, truck drivers, manufacturing workers, and public employees throughout the U.S. and Canada. Roger Philpot, "Who They Are, Where They Work," *Kentucky Coal Miners, Where They Worked, Where They Lived*, 2012, www.coal-miners-in-Kentucky.com.

2. The Pocahontas Coalfield (Flat-Top Pocahontas Coalfield) is located in eastern Mercer County, McDowell County, and Wyoming County in West Virginia, and Tazewell County, Virginia. The Pocahontas Field was mainly the domain of the Norfolk & Western Railway. Mining began in the Pocahontas Coalfield in the 1880's around Bramwell and nearby Pocahontas, Virginia, and really accelerated when the Norfolk and Western built their "Ohio Division" from Bluefield to Kenova, West Virginia, in the 1890's. A large number of European immigrants and African-Americans from the Deep South came to the area to work in the mines. Coalfields of the Appalachian Mountains, www.coalcampusa.com.

3. The Hungarian Revolution of Kossuth in 1848 was a Hungarian War of Independence and a fight for liberation and democracy, and March 15, Hungarian National Day, is the day that commemorates this 1848 War of Independence and Fight for Liberation and Democracy. Lajos (Louis) Kossuth was Governor of Hungary and leader during the Fight for Liberation and Democracy which eventually was defeated by the Austrian

Habsburg and Russian Czarist Armies in 1848–1849. Kossuth envisioned a Federation in the Kingdom of Hungary in which all nationalities participated in a vibrant democratic system based on fundamental democratic principles such as equality and parliamentary representation. This bloody War of Independence eventually led to a compromise known as the "Austro-Hungarian Empire" in which Hungary gained some autonomy. However, Kossuth would have no part of the new empire and demanded full independence until his death. The War of Independence and Fight for Democracy also set the seeds for Hungary's dismemberment after WW I. The American Hungarian Federation, www.americanhungarianfederation.org.

4. The *Lusitania* was built in 1906 for the Cunard Steamship Co., Ltd. It was a 1,550 gross ton ship, length 762.2ft x beam 87.8ft, four funnels, two masts, four screws, and a speed of 25 knots. The *Lusitania* provided accommodations for 563-1st, 464-2nd, and 1,138-3rd class passengers. It launched on June 7, 1906, from Liverpool to Queenstown (Cobh) and New York. It broke both eastbound and westbound records for the fastest passage. The *Lusitania's* last Liverpool-New York voyage started on April 17, 1915. It sailed from New York on May 1, 1915, and was torpedoed and sunk by the German submarine U.20 on May 7, 1915, near the Head of Kinsale, Ireland. As a result of the torpedo attack, 1,198 lives were lost. NRP Bonsor, *North Atlantic Seaway*, vol.3, p.913, on The Ships List, "Lusitania 1906," www.theshipslist.com.

5. The League of Nations was an international organization headquartered in Geneva, Switzerland, and created after the First World War to provide a forum for resolving international disputes. Though first proposed by President Woodrow Wilson as part of his Fourteen Points Plan for an equitable peace in Europe, the United States never became a member of the League of Nations. "Milestones: 1914–1920, The League of Nations, 1920." U.S. Department of State, Office of the Historian, history.state.gov.

6. George E. Creel was a journalist, politician, and author. As chairman of the Committee on Public Information (CPI) during WW I, Creel sought to influence public opinion and gain support at home and abroad for the war effort. After WW I ended in 1918, the CPI was disbanded, and Creel returned to private life. Kimberly Harper, *George Creel (1876–1953)*, on The State Historical Society of Missouri, Historic Missourians, shsmo.org.

14. I Become a Coal Operator

1. An article enhancing the Himler Coal Company information supplied by Martin Himler in Chapter 14, "I Become A Coal Operator," of *The Making Of An American (The Autobiography of a Hungarian Immigrant, Appalachian Entrepreneur, and OSS Officer)* is: Doug Cantrell, "Himlerville: Hungarian Cooperative Mining in Kentucky," *The Filson Club History Quarterly* No.4, October (1920): 513–542, The Filson Historical Society, filsonhistorical.org. Article also on file with editor.

2. The Federal Fuel Administration was established as emergency Executive Order 2690, August 23, 1917, to regulate the production, distribution, and consumption of

coal, coke, natural gas and petroleum products. The Administration was abolished on June 30, 1919. National Archives, Records of the U.S. Fuel Administration [USFA], (Record Group 67), 197–20, 943cu.ft., www.archives.gov.

3. Armistice Day, November 11, formerly observed in the United States in commemoration of the signing of the armistice ending World War I in 1918. Since 1954, Armistice Day has been incorporated into the observances of Veterans Day. *The American Heritage Dictionary of the English Language*, 5th ed. (Boston: Houghton Mifflin Harcourt, 2011), 97.

4. Following the Armistice that ended WW I, Hungary was occupied by enemy forces that paved the way for a Communist coup carried out mostly by non-Hungarians in 1919. This coup led to the territorial dismemberment of Hungary at the Treaty of Trianon in 1920. Hungary lost nearly three-fourths of her territory, and millions of Hungarians were forced under the oppressive rules of newly-created neighboring states. "Hungarian Historical Chronology," Hunmagyar.org, www.hunmagyar.org.

5. "Gemütlich" (guh-moot-lik) is a German adjective meaning comfortable, pleasant, cozy, friendly. Dictionary.com, www.dictionary.reference.com.

6. The Bolshevik Regime is a left-wing majority group of the Russian Social Democratic Workers' Party that adopted Lenin's theses on party organization in 1903. The Bolshevik Regime seized power in Russia in November, 1917. *The American Heritage Dictionary of the English Language*, 5th ed (Boston: Houghton Mifflin Harcourt, 2011), 207.

7. Miklós Horthy de Nagybánya was a Hungarian admiral and statesman who served as Regent of Hungary from 1920–1944. Following World War I, Admiral Horthy organized a counter-revolution against the Communist government of Béla Kun, and Horthy was made Commander in Chief of the Hungarian armed forces in 1919. With Horthy's election as Regent in 1920, Hungary became the first post-World War I nationalist dictatorship in Europe and ruthlessly suppressed all political opposition. During the 1930's, Horthy's government unsuccessfully wanted the Treaty of Trianon, Hungary's settlement with the Allies in 1920, to be revised to return parts of Czechoslovakia, Romania, and Yugoslavia. In 1938–1940, under German/Italian arbitrage, the first and second Vienna Awards gave Hungary sections of Slovakia and Romania. In return for the Romanian territory, Hungary agreed to fight with Germany and Italy during World War II. However, when defeat of the Axis powers seemed imminent in 1942, Horthy began negotiating a separate peace treaty with the Allies. When 1944 brought the German occupation of Hungary, Horthy chose to install a pro-German government rather than face German takeover. When the USSR invaded Hungary in 1944, Horthy tried to surrender to them, but the Germans heard about the impending surrender and arrested Horthy. At the end of WW II, Horthy was captured by the United States Army and was held in protective custody until the end of 1945. Horthy spent the rest of his life in Portugal. The World War Biography, worldatwar.net.

8. Admiral Miklós Horthy de Nagybánya, who led the counterrevolutionaries and their government, entered Budapest and was appointed Regent and Head of State

of Hungary in March, 1920. Horthy restored the monarchy, albeit separated from Austria, and thwarted two attempts, (March and October, 1921), by former Austrian Emperor and King of Hungary Charles I to regain the Hungarian throne. A militantly anti-communist authoritarian government composed of military officers entered Budapest on the heels of Horthy and the counterrevolutionary Romanians. A "white terror" then ensued that led to the imprisonment, torture, and execution without trial of Communists, Socialists, Jews, leftist intellectuals, sympathizers with the Karolyi and Kun Regimes, and others who threatened the traditional Hungarian political order that Horthy and his officers sought to re-establish. Approximately 5,000 people were executed, 75,000 others jailed. The "white terror" forced nearly 100,000 people to leave Hungary, most of them Socialists, intellectuals, and middle-class Jews. "'White Terror' In Hungary, 1919–1921." *On War*, www.onwar.com.

15. Himlerville, Kentucky

1. A horse that is used for riding or driving a hackney. *The American Heritage Dictionary of the English Language*, 5th ed (Boston: Houghton Mifflin Harcourt, 2011), 788.

2. An article supporting the Himler Coal Company and Himlerville information supplied by Martin Himler in Chapter 15, "Himlerville, Kentucky," of *The Making of an American (The Autobiography of a Hungarian Immigrant, Appalachian Entrepreneur, and OSS Officer)* is: "Co-operative Mining Enterprise a Success," *Saward's Journal*, August 27, 1921, 359. Article on file with editor.

3. The Ruhr is a region of northwest Germany along and north of the Ruhr River, which flows about 235 km (145 mi) westward to the Rhine River near Duisburg. The industrial development of the region began in the 1800's. *The American Heritage Dictionary of the English Language*, 5th ed (Boston: Houghton Mifflin Harcourt, 2011), 1533.

4. "Four Men, Two Generations, One Enduring Institution." "The father starts the business, and eventually the son steps up to succeed him. It's the storyline behind generations of American economic growth and expansions… The Huntington story is that plot, except tripled: Pelatiah Webster Huntington had not one son, but three, who in succession carried their father's vision forward." In the words of Clair E. Fultz, former CEO Huntington National Bank, "F.R. or Franz as he was also known, was P.W.'s middle son to whom P.W. turned when P.W. was ready to relinquish the Huntington National Bank's presidency in 1914. The choice made sense. Though Theodore Sollace (T.S.) Ropes was older, he suffered from tuberculosis throughout his life, and poor health would eventually cause him to reduce his responsibilities at the bank. Franz, however, was a bulldog of a man, shorter than his father, but aggressive and energetic with a commanding voice and a presence. He also was an accomplished painter, and when he looked around him at the Scioto River Valley, his artist's imagination took hold. He could see railroads, manufacturing plants, and mines where at the time, there

was only undeveloped countryside. Franz had the foresight to develop a bond department at Huntington. In 1923, he led the bank through two mergers, just a few weeks apart, which effectively created Huntington's first savings department." Francis Ropes Huntington died of heart disease at the age of 52, and his older brother, Theodore Sollace Ropes, succeeded him as President of Huntington National Bank. Geoffrey C. Wright and Richard Sylla, *Genealogy of American Finance* (New York: Columbia University Press, 2015), 162, Clair E. Fultz, *Huntington: A Family & A Bank* (Columbus: Huntington Bancshares Incorporated, 1989), 48, Clair E. Fultz, "The Huntington: A Story of the Huntington National Bank of Columbus." Delivered at the "1966 Columbus Dinner" of the Newcomen Society in North America (Columbus, Ohio, May 25, 1966), 19, website "for Huntington Bancshares, Inc.," built in 2016 by Clique Studios. www.huntington150years.com.

5. Norddeutscher Lloyd was founded in 1857 and grew to be one of the largest and most successful shipping companies in the world. The Norddeutscher Lloyd used only steamships for overseas transportation and promoted emigration through advertising campaigns in several European countries. Deutsche Auswanderer—Denmark, deutsche-auswanderer-datenbank.de.

6. The *Rotterdam* was a 24,149 gross ton ship, length 650.5ft x beam 77.4ft, two funnels, two masts, twin screw, and a speed of 16 knots. The *Rotterdam* accommodated 530-1st, 555-2nd, and 2,124-3rd class passengers. In 1908, the ship made its first voyage from Rotterdam, Netherlands, to New York, and its last Rotterdam-New York-Rotterdam voyage was 1939. NRP Bonsor, *North Atlantic Seaway*, vol.3, p.913, on The Ships List, Rotterdam 1903. www.theshipslist.com.

7. Pápa is a small, beautiful town in Veszprém County in Hungary near Gyor (45km) and Veszprém (55 km) at the borders of the Bakony Mountains and the Little Hungarian Plains. István Molnár, "Overview," Pápa (Veszprem, Hungary), *Pápa Város*, July 8, 2000. www.crwflags.com.

8. Nazism under the leadership of Adolf Hitler, (1889-1945), also is known as the National Socialist German Workers' Party or Nazi Party and grew into a mass movement that ruled Germany through totalitarian means from 1933 to 1945. Founded in 1919 as the German Workers' Party, the Nazism group promoted German pride and anti-Semitism. The group also expressed dissatisfaction with the terms of the Treaty of Versailles, the 1919 peace settlement that ended WW I, (1914-1918), and required Germany to make numerous concessions and reparations. "Nazi Party," History, www.history.com.

16. Life in Himlerville

1. The Congressional career of Katherine Gudger Langley illustrates a highly unusual route to Congress. Her husband, John Langley, resigned his House of Representatives seat after being convicted of violating Prohibition laws by trying to sell

1,400 bottles of whiskey. While John Langley's conviction was being appealed, John ran for his Congressional seat and won re-election in 1924. When the U.S. Supreme Court refused to overturn the decision, John Langley resigned from the 69th Congress (1925-1927) and was sentenced to the federal penitentiary in Atlanta for two years. Katherine Langley, as a result of her husband's conviction, was socially ostracized in the conservative social scene in Washington, D.C. Katherine Langley resolved to clear her husband's name by running for his seat in the 70th (1927-1929) Congress. With John's help from prison, she won election to Congress. Katherine also won a second term, but was defeated in her third attempt for re-election in 1930. Congresswoman Langley retired to Pikeville, Kentucky, where John Langley had earlier resumed his law practice. John Langley died in January, 1932, of pneumonia, still arguing that he had been sent to prison unjustly. Katherine Langley served as a postmistress and was twice elected as a district railroad commissioner. Katherine Langley died in Pikeville, Kentucky, on August 15, 1948. "History, Art, and Archives," United States House of Representatives, house.gov.

2. The *Survey Graphic*, created in 1921 by Paul Kellogg, is a companion journal to the *Survey*. The *Survey Graphic* was aimed less at professional social workers and more generally at socially conscious members of the public. Kellogg wanted the *Survey Graphic* to represent "social fact" and not "opinion." He did not want to tell people what they should think, but wanted instead to "provoke citizens everywhere into an awareness of new programs for social reform," and Kellogg hoped to get citizens to see for themselves the necessity for particular social change. "Viewed with the hindsight of history, *Survey Graphic* in the 1920's produced analyses of social and political issues that were years ahead of their time." Cara Finnegan, "The Paul Kellogg Era: Social Reform in The Survey and Survey Graphic," *Social Welfare and Visual Politics, The Story of Survey Graphic*, newdeal.feri.org.

3. Mary Domosley Koblass and her family came to America from Hungary in 1910, and the Domosleys moved to the "Promised Land" of Himlerville in 1919. Mary grew up in Himlerville, graduated in 1928 from Magnolia High School in Matewan, West Virginia, and married Charles Koblass, a native of Poland and a hotel chef, in 1929. Mary received her Bachelor of Science degree from Pikeville College in 1959, and she taught in Martin County and in Mingo and Wayne Counties in West Virginia. Mary is the author of a series of articles, "Himlerville Revisited," that appeared in *The Martin Countian* from December 3, 1975–July 28, 1976. *The Martin Countian* was then the local newspaper serving Martin County, and the "Himlerville Revisited" articles were published in recognition of America's Bicentennial. Mary's "Himlerville Revisited" articles are a factual, detailed, and delightful account of life in Himlerville, and the articles include photographs of Himlerville, menus from dinners held at Martin Himler's home, and poems composed by Mary. Mary Domosley Koblass, "Himlerville Revisited," *The Martin Countian*, December 3, 1975–July 28, 1976. Articles on file with editor.

17. The Lost Dream

1. William Edgar Borah, a prominent Republican attorney, fierce Progressive, and U.S. Senator from Idaho from 1907 until 1940, was known for his public speaking skills and his advocacy for peace and the outlawry of war. Borah was Prosecuting Attorney in the nationally-publicized trial of three labor union officials accused of murdering former Idaho Governor Frank Steunenberg. "William Borah, U.S. Senator," Geni, www.uidaho.edu.

2. George William Norris was a liberal Progressive Republican congressman from Nebraska whom many consider to be history's greatest United States Senator. Norris served in the Senate from 1913–1943, and his greatest legislative monument is the Tennessee Valley Authority. United States Senate, www.senate.gov.

3. Henry Agard Wallace is the thirty-third Vice-President of the U.S. (1941–1945) who also served as the U.S. Secretary of Agriculture (1933–1946) and the U.S. Secretary of Commerce (1945–1946). Wallace once declared that his greatest aspiration was "to make the world safe for corn breeders." As Secretary of Commerce, Wallace feuded bitterly with President Harry S. Truman over America's confrontational posturing with the Soviet Union, and the clash earned Wallace a reputation among his detractors as a "Stalinist Stooge." Wallace's 1948 run for the Presidency as the Progressive Party candidate has been termed "the closest the Soviet Union ever came to actually choosing a President of the United States" because Wallace advocated closer cooperation with the Soviet Union. Alex Altman, "America's Worst Vice Presidents," *Time Online*, content.time.com.

4. A bohemian is a person with artistic or literary interests who disregards conventional standards of behavior. *The American Heritage Dictionary of the English Language*, 5th ed (Boston: Houghton Mifflin Harcourt, 2011), 206.

5. The *Nation* was founded by abolitionists in 1865. The *Nation* is America's oldest weekly magazine covering American political and cultural life and is considered the "flagship" of the political Left. The *Nation* headquarters is located at 33 Irving Place, New York, New York 10003. www.thenation.com.

6. The *New Republic* was founded in 1914 as a journal of opinion and champion of progressive ideas. Chris Hughes, a Facebook founder and owner of *New Republic*, sold *New Republic* to Win McCormack in February, 2016. McCormack is a publisher and editor based in New York and Portland, Oregon, and McCormack founded the literary quarterly, *Tin House*. McCormack appointed Hamilton Fish, publisher of the *Washington Spectator* and a former publisher of the *Nation* to be *New Republic* publisher and editorial director. newrepublic.com, and Ravi Somaiya, "A Buyer Is Found for the *New Republic*," *New York Times*, February 27, 2016:B3, New York edition.

7. Linotype is a trademark for a machine that sets type on a metal slug operated by a keyboard. *The American Heritage Dictionary of the English Language*, 5th ed (Boston: Houghton Mifflin Harcourt, 2011), 1022.

8. President Franklin D. Roosevelt took legislative action in 1933 to bring about immediate economic relief as the result of the Great Depression, and the relief was called the New Deal. The New Deal included several agencies such as the Works Progress Administration (WPA), Civilian Conservation Corps (CCC), National Recovery Administration (NRA), Federal Deposit Insurance Corporation (FDIC), Tennessee Valley Authority (TVA), and the Securities and Exchange Commission (SEC). "The New Deal." United States History, www.u-s-history.com.

9. Adolph Joachim Sabath was a Democratic congressman from Chicago, Illinois, who was elected to the Sixtieth and to the succeeding twenty-three Congresses. Congressman Sabbath immigrated to the U.S. from Czechoslovakia at the age of 15, and he was a leading opponent of Prohibition in the 1920's. Burton A. Bokerman, "SABATH, Adolph Joachim, (1866–1952)," Biographical Directory of the United States Congress 1774–Present, bioguide.congress.gov.

10. To deal with America's continued unemployment problems, Franklin D. Roosevelt launched a second, more aggressive series of federal programs to assist labor, a series often called the Second New Deal, in the Spring of 1935. Also in 1935, the National Labor Relations Act, also known as the Wagner Act, created a board to supervise union elections and prevent businesses from treating their workers unfairly. "New Deal (Second New Deal)." History, www.history.com.

11. Joseph Stalin, (1879-1953), was born in the Russian peasant village of Gori, Georgia, as Vissarionovich Dzhugashvili and was later known as Joseph Stalin. Stalin rose to power as General Secretary of the Communist Party in 1922 and became a Soviet dictator upon Vladimir Lenin's death in 1924. Stalin understood that "cadres are everything": if you control the personnel, you control the organizations. Stalin controlled all appointments, set agendas, and moved around party staff in such a way that eventually everyone who counted for anything owed their position to Stalin. The purges of opposition to Stalin extended down into every local Party cell, and purges targeted nearly all of the intellectual professionals. Stalin turned the Soviet Union from a backward country into a world superpower at unimaginable human cost. "Joseph Stalin (1879-1953)," *REDFiles*, Copyright 1999 by Abamedia unless otherwise noted, www.pbs.org, and biography.com editors, "Joseph Stalin Dictator (1879-1953)," Biography, A&E Television Networks, July 29, 2015, www.biography.com.

12. Gerald Lyman Kenneth Smith was a Disciples of Christ minister who built a series of tourist attractions with a religious theme in Eureka Springs, Arkansas. Smith was known for far-right activism, particularly for anti-Semitic and Fascist causes. Smith once considered joining the Silver Shirts, a pro-Nazi group led by William Dudley Pelley. Smith had several unsuccessful runs for the U.S. Senate and the Presidency, and his last run for the Presidency was as a candidate for the Christian Nationalist Party in 1956. Michael Guager and Glen Jeansonne, "Gerald Lyman Kenneth Smith (1898–1976)," The Encyclopedia of Arkansas History and Culture, Updated November 15, 2016, www.encyclopediaofarkansas.net.

13. The National Industrial Recovery Act of 1933 was one of the most important and daring measures of President Franklin D. Roosevelts's New Deal. The legislation was enacted in the first 100 days of Roosevelt's first term in office and was his initial effort to reverse the effects of the Great Depression with a Public Works Program. The National Industrial Recovery Act was to remain in force for two years with the goal to alleviate unemployment and stimulate industrial output through government regulation. The United Mine Workers of America quadrupled its membership from 100,000 to 400,000 in less than a year after the Act's passage. In 1935, less than three weeks before the Act would have expired, the U.S. Supreme Court ruled it unconstitutional. legal-dictionary.thefreedictionary.com.

14. John Llewellyn Lewis was President of the United Mine workers of America from 1920 until 1960 and founding President of the Congress of Industrial Organizations (CIO). Lewis was the dominant voice shaping the labor movement in the 1930's. The CIO owed its existence in large measure to Lewis who was a tireless and effective advocate of industrial unionism and of government assistance in organizing basic industry. "John L. Lewis (1880–1969)," AFL-CIO America's Unions, www.aflcio.org.

18. The American Miner

1. The Taft-Hartley Act, passed in 1947, remains the cornerstone of today's United States labor law. The Act amended the Wagner Act of 1935 in order to balance the power given to labor and the post-World War II attitudes of labor/management relations. "TAFT-HARTLEY ACT," Reference for Business Encyclopedia, www.reference forbusiness.com

2. Muck-raking is searching for and exposing political misconduct in public life. *The American Heritage Dictionary of the English Language*, 5th ed (Boston: Houghton Mifflin Harcourt, 2011), 1154.

3. James Caesar Petrillo was "beloved by some, despised by many." Petrillo was one of the most colorful, powerful, and controversial leaders of the twentieth century. Petrillo was President of the United Federation of Musicians and ruled the organization with an iron fist from 1940 to 1968. "Chicago Stories," WTTW, www.wttw.com.

4. William Morris (Willie) Bioff was an American organized crime figure who operated as a labor leader in the movie production business from the 1920's through the 1940's. Bioff extorted millions of dollars from movie studios with the threat of mass work stoppages, and on November. 4, 1955, "he was blown to bits at the wheel of his car—payback from the boys in the Windy City." Jack Backstreet, "Willie Morris Bioff," IMDb, www.imdb.com.

5. The American Hungarian Federation (AHF), founded in 1906 in Cleveland, Ohio, is the largest Hungarian umbrella organization and among the oldest ethnic organizations in the United States. The AHF is a 501(c)(3) organization that strives to retain Hungarian heritage and cultural preservation and to communicate AHF's

message to the Western media and to the world. Since the AHF's founding, the organization has been a stalwart supporter of freedom both at home and abroad. www.americanhungarianfederation.org.

6. The World Federation of Hungarians, the Magyarok Világszövetsége, is the largest Hungarian non-government organization and was founded in 1938. The Magyarok Világszövetsége has three parts: Hungary Region, Carpathian Basin Region, and Hungarians in the West Region. www.crwflags.com.

7. The Anschluss is the political union of Austria with Germany and was achieved through annexation of Austria into a greater Germany by Adolf Hitler in 1938. History, www.history.com.

8. Reich is the territory or government of a German, or First Reich, from 962 to 1806; the German Empire, or Second Reich, from 1871 to 1919; the Weimer Republic, from 1919 to 1936; or the Third Reich, from 1933 to 1945. *The American Heritage Dictionary of the English Language*, 5th ed (Boston: Houghton Mifflin Harcourt, 2011), 1481.

9. Béla Imrédy was a Catholic financier who steered Hungarian economic policy in the chaotic 1930s. Béla Imrédy was appointed Prime Minister by Regent Miklós Horthy in 1938. In 1939, Prime Minister Imrédy was forced out of office on revelations that he had Jewish ancestry. This did not moderate Imrédy's anti-Semitic views; he returned as the head of a new Fascist party and almost became Prime Minister under the German occupation in 1944. A "Peoples Tribunal" convicted Imrédy after the war for war crimes and Nazi collaboration. Béla Imrédy was executed by firing squad on February 28, 1946. "1946: Bela Imredy, Hungarian Fascist Prime Minister," Executed Today, www.executedtoday.com.

10. "The Smallholder Party was the key anti-Nazi and anti-Communist force in Hungary. Unfortunately, it [the Party] no longer exists." Electronic mail conversation with Hungarian-American Charles Fenyvesi, internationally recognized author and retired *Washington Post* and *U.S. News and World Report* columnist, October 17, 2015.

11. Béla Kun, (1886–1936), the Hungarian revolutionary, was responsible for founding the world's second Communist government when the disintegration of the Austro-Hungarian empire came at the close of WW I. The founder of the Hungarian Communist Party, Kun served in the Austro-Hungarian Army in 1914 and was taken prisoner by the Russians in 1916. Kun played no further part in the war until the Russian Revolution and the ascent of the Bolsheviks to power. Kun returned to Hungary as an ally of the Bolsheviks, and Kun's intention was to stir a revolutionary uprising in Hungary. Kun was successful with his revolution plan until unpopular decisions were made to nationalize industry and agriculture. Kun retained power through the ruthless use of the armed forces, but opposition to Kun's government grew, and on August 1, 1919, Kun's government fell during invasions from both the Czechs and Romanians and the French-sponsored counter-revolutionary force led by Admiral Miklós Horthy de Nagybánya (which succeeded in establishing Horthy in the Hungarian government).

Kun fled to the Soviet Union, became a party operator within Stalin's government, and is believed to have been killed in Stalin's innumerable purges. Michael Duffy, "Who's Who–Bela Kun," Saturday, August 22, 2009, www.firstworldwar.com.

12. Draza Mihailovic was a Yugoslav Serb General who organized his Serbian forces to resist the Nazi invasions. Mihailovic was recognized as an ally by the U.S. in World War II, and Mihailovic was betrayed and executed at the hands of Nazi-supporting Yugoslav Communists in Autumn, 1941. Congressional Record, reprinted from Congressional Record, Washington, D.C., Thursday, November. 19, 1987, www.srpska-mreza.com.

13. Josip Broz Tito is a former President, (1953–1980), and Prime Minister, (1944–1963), of Yugoslavia who was born Josip Broz on May, 1892, and adopted the name Tito, common in his home region, for conspiratorial cover in the 1930's. Tito was a strong Communist, and when the Germans attacked the Soviet Union on June 22, 1941, Tito formed guerrilla units to fight the occupation forces of Germany, Italy, Hungary, and Bulgaria that had sliced up Yugoslavia like a cake. Tito was known as a man of stubborn courage, ready to fight and intrigue and endure hardship and risk death for his beliefs. Raymond H. Anderson, "Giant Among Communists, Governed Like a Monarch," *New York Times Online,* May 5, 1980, graphics8.nytimes.com.

14. "Cairo is the probable location where Colonel Himler began his Office of Strategic Services work abroad, working on MO (Morale Operations) and editing/composing propaganda to be dropped over Hungary." Electronic mail conversation with Duncan Bare, Ph.D. Candidate and specialist in Hungarian-American Intelligence Studies, University of Graz, Austria, Monday, June 1, 2015.

15. The Flying Fortress, the Boeing B-17 heavy bomber, served in almost every theater of WW II. When Hitler's forces invaded Poland in September, 1939, there were only thirteen operational Flying Fortresses "and the United States was the only country with a strategic bomber when the war began." Larry Dwyer, "Boeing B-17 Flying Fortress," The Aviation History Online Museum, site created September 8, 1996, updated December 9, 2015, www.aviation-history.com.

16. A lorry is a large, heavy motor vehicle for transporting goods or troops; a truck. www.oxforddictionaries.com.

17. The Grand Mufti is the highest official of religious law in a Suni or Ibadi Muslim country. The Grand Mufti issues legal opinions and edicts, fatwa, on interpretations of Islamic jurisprudence for private clients or to assist judges in deciding cases. "WHAT DOES GRAND MUFTI MEAN IN ENGLISH," *English Dictionary Online,* englishdictionary.edu.

18. Erwin Rommel was a German commander who was known as "Desert Fox" for his North African campaign of pushing the British back hundreds of miles. Finally, in October 1942, the numerically superior British halted Rommel's advance near El Alamein, Egypt. Running low on tanks, ammunition, and fuel, Rommel retreated. Hitler sent Rommel a letter telling him not to yield "even a yard of ground." "As to

your troops," the führer added, "You can show them no other road than that to victory or death." Rommel disobeyed for fear his force would be completely annihilated. Two months later, the Germans were kicked out of North Africa. Jesse Greenspan, "8 Things You Might Not Know About Erwin Rommel," History, October 14, 2014, www.history.com.

19. A burnous is a one-piece hooded cloak that is worn by Arabs and Berbers. *Merriam-Webster,* www.merriam-webster.com.

20. "Bari was an important place in Office of Strategic Services work. Bari was the location where Company B of the 2677th Regiment was located, and Colonel Himler and other intelligence agents collected in Bari under the aegis of SICE (Secret Intelligence Central Europe)." Electronic mail conversation with Duncan Bare, Ph.D. Candidate and specialist in Hungarian-American Intelligence Studies, University of Graz, Austria, Monday, June 1, 2015.

21. Green troops are inexperienced and disadvantaged military troops. Newsok.com.

22. "Rolling stock" is the equipment available for use as transportation, as automotive vehicles, locomotives, or railroad cars, owned by a particular company or carrier. *The American Heritage Dictionary of the English Language,* 5th ed (Boston: Houghton Mifflin Harcourt, 2011), 1521.

23. In May 1945, around 10 million refugees inside the boarders of Germany, Austria, and Italy relied on the Allied Forces for their immediate needs and assistance in resettlement. At the end of the summer of 1945, 1.5 million refugees refused to return to their home countries. These refugees were defined as "Displaced Persons." Milano served as one of the many Displaced Persons camps where Jewish Displaced Persons co-habited with refugees from various other nationalities. "Historical Note," *Guide to the Records of the Displaced Person Camps and Centers in Italy 1945-1955 RG 294.3,* digifindingaids.cjh.org.

24. Caserta, Italy, became the location of Headquarters of the Allied Forces from 1944–1946. Caserta was the scene of the signing of the German surrender on April 29, 1945. "Caserta Center - Allied HeadQuarters and stop for many soldiers," *Rest Camps R&R Centers and Replacement Depots Used by 5th Army,* www.custermen.com.

25. Brenner Pass between Italy and Austria is 59 miles long and is 4,497 feet high at its topmost point. This makes Brenner Pass one of the lowest important Alpine passes. The only way to cross the Alps in the 1940's was through the mountain pass of Brenner. Thus, the Brenner Pass was an important German supply route to Italy. Ernest M. Young, 432nd Squadron—World War II, "Brenner Pass November 17, 1944," *Bomb Group 17,* sites.google.com/site/bombgroup17/brenner.

26. The Black Hand was a secret society organized for acts of terrorism and blackmail and was active in the United States in the early 20th Century. *The American Heritage Dictionary of the English Language,* 5th ed (Boston: Houghton Mifflin Harcourt, 2011), 191.

27. "*Tosca* is one of the most lethal of operas. None of the central characters make it to the end alive, whether they be hero or villain. *Tosca* is fundamentally fiction of the eighteenth century seen through nineteenth century eyes, and twenty-first century eyes are still captivated and enthralled by *Tosca*. *Tosca* is an Italian tale of romance over politics and features a heroic painter, despicable ruler, and an opera superstar, Tosca herself." The arias, or long solos, of *Tosca* are well known by both Americans and Europeans. www.theopera101.com.

28. The carabinieri is the Italian national police force. dictionary.reference.com.

29. Berchtesgaden is a German town in the Bavarian Alps. Adolf Hitler stayed at the mountain retreat area of Obersalzverg located above Berchtesgaden where he wrote the second part of *Mein Kampf*. www.thirdreichruins.com.

30. Justice Robert Houghwaut Jackson is a former Associate Justice of the U.S. Supreme Court, a former U.S. Attorney General, and a former U.S. Solicitor General who is the only person in U.S. history to have held all three of these offices. Justice Jackson was a diehard Democrat who became a friend and advisor to Franklin D. Roosevelt during Roosevelt's run for the governorship of New York. President Harry S. Truman appointed Jackson as chief counsel for the Nuremberg Trials, and Jackson was the chief prosecutor for infamous Nazis such as Hermann Göring and Rudolph Hess. biography.com editors, "Robert H. Jackson Biography," Biography, www.biography.com.

31. General William Joseph "Wild Bill" Donovan was the Head of the Office of Strategic Services, the precursor to the Central Intelligence Agency, during World War II. Donovan is the former recipient of the Medal of Honor, the Distinguished Service Cross, the Distinguished Service Medal, and the National Security Medal, and the only American to have received these four highest awards of the U.S. During World War II, Donovan founded and led the Office of Strategic Services, and Donovan served as an assistant to Robert Jackson, Chief American Prosecutor at the Nuremberg War Crimes Trials. Donovan also served as the U.S. Ambassador to Thailand in 1953. Michael Robert Patterson, "William Joseph Donovan Major General United States Army," Arlington National Cemetery, www.arlingtoncemetery.net.

32. Colonel John Harlan Amen was Chief of the Interrogation Division, Office of Chief of Counsel United States Army, and served as Nuremberg Prison Chief Interrogator during the Nuremberg War Trials. Amen was the second husband of Marion Cleveland Deli, the daughter of former President Grover Cleveland. "John Harlan Amen," Geni, www.geni.com.

33. Nuremberg, Germany, was the location of the World War II 1945–1949 trials held for the purpose of bringing Nazi war criminals to justice. The defendants included Nazi Party officials and high-ranking military officers, along with German industrialists, lawyers, and doctors. All were indicted on such charges as crimes against peace and crimes against humanity. Nazi leader Adolf Hitler (1889–1945) committed suicide and never was brought to Nuremberg trial. History, www.history.com.

19. Playing God

1. Dachau, one of the first concentration camps established by the Nazis, was located in the small town of Dachau, approximately 10 miles northwest of Munich. The camp opened in 1933 and closed in 1945. www.holocaustresearchproject.org.

2. Martin Himler was responsible for the apprehension, arrest, and holding of more than 300 Hungarian war criminals and for the interrogation of more than 40 of these "beasts." According to Duncan Bare, Ph.D. Candidate and specialist in Hungarian-American Intelligence Studies at University of Graz, Austria, "Martin's [Office of Strategic Services] job was gathering intelligence on all things Hungarian and overseeing a handful of other men who did the same thing. Himler's reporting spanned the breadth of Hungarian affairs, including political, economic, and social info of all kinds. Martin was a pre-networker and had an absolutely amazing web of friends, acquaintances, and contacts in Hungary, particularly among the 'good guys' or those members of the Hungarian Independence Movement (MFM), the Smallholder Party, and other liberal-democratic groups." "Colonel Himler's OSS code name for most of 1944/45 was "Dean." Himler arrived in Budapest in 1945 from Camp Marcus W. Orr, near Salzburg, Austria, with the first batch of war criminals." Their arrival in Budapest is documented on http://filmhiradokonline.hu/watch.php?id=5940. Electronic mail conversation with Duncan Bare, Ph.D. Candidate and specialist in Hungarian-American Intelligence Studies at University of Graz, Austria, May 30, 2015. Duncan Bare also shared the link http://filmhiradokonline.hu/watch.php?id=5940, to the filming of the return to Budapest of some of the criminals interrogated by Colonel Himler at Camp Orr, Salzburg, Austria. Electronic mail conversation with Duncan Bare, May 31, 2015.

3. *The Grand Duchess of Gérolstein* is one of a series of French operettas by Jacques Offenbach. This operetta was first performed at the Théâtre de Variétés in Paris on April 12, 1867, *and The Grand Duchess of Gérolstein* tells the hilarious tale of a sultry grande dame who rules the tiny Duchy of Gérolstein. *The Grand Duchess of Gérolstein* is filled with colorful and amusing characters, much comedy, and irresistible music. *The Grand Duchess of Gérolstein, The Guide to Light Opera & Operetta,* www.musical theatreguide.com.

4. The American Jewish Joint Distribution Committee (JDC), colloquially called "The Joint," is a world-wide Jewish relief organization that was established during World War I in 1914. JDC's relief activities, emigration aid, and rescue operations were critical following the Nazi rise to power and the outbreak of World War II. "History of JDC," American Jewish Joint Distribution Community, www.jdc.org.

5. Ferenc Szálasi was the leader and all-powerful head of the Fascist Arrow Cross Movement, the regime that came to power in Hungary with the armed assistance of the Germans on October 15–16, 1944. After this date, the fate of hundreds of thousands of Jews was in Szálasi's hands. Szálasi professed to be a good Christian and a Catholic, and he argued that anti-Semitism was taught in the Bible itself. Szálasi was head of the

state in Hungary during the final three months of Hungary's participation in World War II, after Germany occupied Hungary and removed Miklós Horthy by force. During Szalási's brief rule, his men murdered 10,000–15,000 Jews. When World War II ended, Szálasi was captured by American troops, returned to Hungary, tried by the Peoples' Tribunal in Budapest in open sessions, sentenced to death for war crimes and high treason, and hanged in Budapest on March 12, 1946. "Hungarian Prime Minister Ferenc Szálasi is given the last rites before being hanged as a collaborator, 1946," Rarehistoricalphotos.com, and Charles Fenyvesi, "Two Men, One Evil," *The Washington Post*, December 16, 2003, Final Edition: Editorial;A37. Article on file with editor.

6. A Papal Nuncio is a papal ambassador, a permanent diplomatic representative of the Holy Sea to a state or an international organization. The Papal Nuncio is a permanent representative of the Pope and is assigned with the task of safeguarding the interests of the Holy Sea. "What is a Papal Nuncio?" Vatican, June 5, 2013, vatican.com.

7. *The Debacle* is the nineteenth century novel in the French novelist Émile Zola's, (1840–1902), great Rougon-Macquart cycle and is the greatest of all war novels. *The Debacle* is a forceful and deeply moving tale of close friendship formed as a result of war, and *The Debacle* also is a fascinating chronicle of the events that led, in Zola's words, to "the murder of a nation." Zola was an exemplar of the literary school of Naturalism, and he was intensely involved in the political liberalization of France. OverDrive, www.overdrive.com.

8. Prime Minister Döme Sztójay (1883–1946) was a pro-Nazi military leader in Hungary. Germany occupied Hungary on March 19, 1944 and three days later, Sztójay was appointed prime minister and foreign minister of the puppet Hungarian government installed by the Nazis. Under Sztójay's rule, Hungarian authorities issued anti-Jewish legislation, put Jews in ghettos, pillaged their property, and deported them to Nazi extermination camps. As the Soviet army approached Budapest, Sztójay fled Hungary with the Nazis. He was caught by American troops who returned him to Hungary in October, 1945. On March 22, 1946, Sztójay was found guilty of war crimes and crimes against the people by Hungarian court, was sentenced to death, and shot. www.yadvashem.org.

9. The six-pointed Jewish Star, the Shield of David, is a hexagram that is the most common and recognized sign of Judaism and Jewish identity. www.menorah.org.

10. Knight Béla Miklós de Dálnok, (June 11, 1890–November 21, 1948), was a Hungarian military officer and politician who served as Prime Minister of Hungary from 1944 to 1945. Miklós was born in Budapest and was appointed military attaché to Berlin and Stockholm between 1933–36. Miklós became commanding general of the Hungarian First Army on August 1, 1944, and he supported leaving the Axis Powers and the Red Army in World War II. Through Soviet request, Miklós spoke on the radio and made a plea for commanding officers of his Hungarian First Army to defect with their units to the Soviets. This was not successful and only one general defected; he was caught by the Germans and executed. Béla Miklós died in Budapest in 1948

and did not receive military honors at his funeral. Project Gutenberg Self-Publishing Press, www.gutenberg.us.

20. In Hungary

1. Buda is a Hungarian town built on the higher river terraces and hills of the western side of Budapest. Pest spreads out on a flat and featureless sand plain on the Danube River's opposite bank. Budapest, the capital of Hungary, was created in August, 1872, when Pest, Buda, and Óbuda were united into a single municipal borough comprising 10 districts. "History," Budapest.com.

2. Buda Mountain, also known as Castle Hill, is the highest mountain in Budapest. "History," Budapest.com.

3. Count Pál Teleki was Prime Minister of Hungary in 1920–21. Teleki was a delegate to the Paris Peace Conference (1919), and left politics to teach geography at Budapest University. Teleki returned to be Prime Minister in 1939. As Prime Minister, he dissolved various Fascist parties, although he allowed anti-Semitic laws to stand. When Germany invaded Hungary in 1941, Teleki was caught between German demands to help against the Yugoslavs (thus breaking his pledge given in the Paris Treaty) and British threats against helping the Germans or even letting the Germans cross the Hungarian border uncontested. Facing these counter pressures, Count Teleki went home after a cabinet meeting, waited for a phone call confirming that German troops had entered Hungarian soil, then put a gun in his mouth and committed suicide on April 3, 1941. Teleki's suicide note was addressed to Horthy, the Hungarian regent, who had, however reluctantly, given the nod to Germany to use Hungary. The note stated, "Your Serene Highness: We broke our word, out of cowardice, with respect to the Treaty of Permanent Peace [with Yugoslavia] outlined in your Mohács speech. The nation feels it, and we have thrown away its honor. We have allied ourselves to scoundrels, since not a single word is true about the alleged atrocities. Not against Hungarians, not even against Germans. We will become body-snatchers! A nation of trash. I did not hold you back. I am guilty." Winston Churchill paid Teleki tribute in Churchill's *Second World War*. "His suicide was a sacrifice to absolve himself and his people from guilt in the German attack upon Yugoslavia. It clears his name before history. He could not stop the march of the German armies nor the consequences." In Teleki's grave side eulogy, his successor, Prime Minister László Bárdossy, referred to Teleki as the "schoolmaster of the nation." "Count Teleki: The Politics of Suicide," Strangehistory.net, February 18, 2015, www.strangehistory.net, and "Pal Teleki (1879–1941): The Life of a Controversial Hungarian Politician," Updated by Balázs Ablonczy, ACADEMIA, www.academia.edu.

4. Vyacheslav Molotov was a Russian statesmen and diplomat who served as Stalin's Prime Minister, among other official titles. Spartacus Educational, spartacus-educational.com.

5. László Bárdossy became Prime Minister of Hungary with the death of Prime Minister Teleki in 1941. Lászlo Bárdossy's elevation to Prime Minister furthered Hungary's partnership with the Axis powers. He joined in the invasion of Yugoslavia and reclaimed a piece of territory Budapest claimed. Bárdossy also tightened Hungary's anti-Semitic laws with Bárdossy's Third Jewish Law, which attempted to cut Jews out of the economic life of Hungary and began deportation of Jews out of Hungary. Bárdossy was arrested after the war, tried as a war criminal by a People's Court for war crimes and Nazi collaboration, and shot by firing squad in 1946. "1946: László Bárdossy, Former Prime Minister," Executed Today, www.executedtoday.com.

6. On March 9, 1942, Prime Minister Bárdossy resigned, and Miklós Kállay was appointed Prime Minister. During Kállay's first year in office, he followed Bárdossy's pro-German line. With the advance of the Allied forces in December, 1942, Hungary lost 100,000 soldiers, and this loss radically changed Kállay's course. He realized Germany would lose the war. Kállay's first move was to resist German demands to curb anti-German groups. He maintained the freedom of opposition to the press, refused to outlaw Leftist Party activities, and, as a result of Kállay's doing this, Hungary remained the only country in Europe which had a Social Democratic Party. Kállay also resisted German demands towards "the solution of the Jewish question." In 1943–44, in spite of the "Jewish Laws," Hungary was the only country in which Jewish people were a part of society and not forced into ghettos. Hitler got wind of Hungary's move toward the Allies, and this led to Germany's occupying Hungary, removing Mikós Kállay, and installing pro-German General Döme Sztójay as Prime Minister. "The Policies of Prime Minister Kállay and the German Occupation of Hungary in March 1944," Hungary in the Mirror of the Western World, *1938–1958*, Gabor Aron Study Group, hungarianhistory.com.

7. Martin Himler's feelings about Hungary and the country's dismal state following the Treaty of Trianon and the end of World War II are echoed by Frank Koszorus Jr., National President of the American Hungarian Federation. Frank Koszorus Jr., "Setting the Record Straight on Hungary's History," Hungary Today, hungarytoday.hu, article reviewed and submitted by Adam Topolansky.

21. Return Home

1. Verhovay Fraternal Insurance Association was founded in Pennsylvania on February 20, 1886, and was chartered as a sick-benefit and burial association. The Association is named for Gyula Verhovay, former member of the Hungarian Parliament. There were twenty-eight association founders and Mihály Pálinkás, from a small Pennsylvania coal town, was the association's first president. The association's membership grew to 52,292 by the mid-1940's. Emmil Lengyel, *Americans From Hungary* (Connecticut: Greenwood Press, 1948, 1974).

2. The Paris Peace Conference, held from July 1946 until February 1947, had delegates from twenty-one nations who met in Paris to decide on peace terms for Germany's

five allies in World War II: Bulgaria, Hungary, Finland, Italy, and Romania. Negotiators from the United States, Soviet Union, United Kingdom, France, and other Allied Powers agreed upon the provisions of the Paris Peace Treaties, and the Treaties were signed in February, 1947. Provisions of the treaties included monetary reparations, territorial adjustments, and political commitments intended to promote democracy and peace. "Paris Peace Treaties," *Oxford Reference Online*, Oxford University Press, www.oxfordreference.com, and *"Moments in U.S. Diplomatic History, The Paris Peace Conference—1946, Association For Diplomatic Studies and Training*, adst.org.

3. A quintal is a unit of weight equal to 100 kilograms (about 220 pounds). *Merriam-Webster Online,* www.merriam-webster.com.

4. Martin Himler, "If I Were the President...," *Magyar Bányászlap, (Hungarian Miners' Journal*), January 2,1958, article reprinted in *Washington Post,* January 16,1958. Article is included in Appendix of this autobiography.

INDEX

Page numbers in **boldface** refer to illustrations.

Abbázia, 17, 266n5
Academy of Commerce, 6, 10–11, 13–14, 60, 265n8
Accra, Africa, 220–21
Air Transport Command, 221, 227
Ajax, West Virginia, 160–61
Ajax mine, 162, 174
Almassy, Count (land owner in Hungary), 3–4, 120
Amen, Harlan John Colonel, 229–32, 288n32
American Hungarian Federation, 212–17, 277n3, 284n5, 292n7
American Legation, 58, 270n1
American Miner, 209–10
Andor, Nicholas, 198
Antal, István, 214, 234
Az Ujsag, 40, 45

Bako, Johnny, 177–82, 190, 211–12, 217–18, 246
Bakron, Eugene, 61, 66, 68–69
Baldwin-Felts Agency, 96, 108, 150, 209, 273n1
Bárdossy, László, 240, 291n3, 292n5
Bari, Italy, 225, 287n20
Battery Place, 20, 22, 63, 75–76
Beauty, Kentucky, 197, 199
Berchtesgaden, 229, 288n29
Black Legion, 38, 269n6

Bluefield, West Virginia, 24, 96, 108, 276n2
Borah, William, 203, 282n1
Brenner Pass, 227, 287n25
Buck Creek, **129**, 172, 197
Buda Mountain, 239, 268n1, 291n2
Budafok, 12
Budapest: Himler in, 10–13, 16–17, 115–18, 121–23, 164–66, 179–81, 183, 237–39, 241–43, 248–50; Himler discusses, 1, 35, 40, 104, 198, 213, 248–50; notes, 268n1, 275n6, 278n8, 289n2, 290n5, 290n8, 290n10, 291n1, 291n2, 292n5

Cairo, 219, 221–26, 229, 252–53, 286n14
Calumet, Michigan, 80, 111, 272n1
Caserta, Italy, 227, 287n24
Chicago, Milwaukee, St. Paul, and Pacific Railroad, 200, 202–4
Chicago Daily News, 188
Cincinnati, Ohio, 30, 69, 174, 185, 268n7
Cleveland, Ohio: Himler in, 47–48, 203, 205, 211, 247; Himler discusses, 30, 45–46, 99, 199, 208, 212; note, 284n5
Cleveland Weekly, 207, 217
Coeburn, Virginia, 143–45
Creel, George, 158, 277n6
Crotty, Dr. (Columbus, OH, cancer specialist), 194

Dachau, 232, 289n1
Debrecen, 120, 275n3
Demjen, Frank, 172

Detroit, Michigan, 30, 111, 120, **135**, 205–7, 212–13, 217
Donovan, J. William General, 229–32, 234, 245, 271n5, 288n31
Drozdy, Victor, 181

Eckhardt, Tibor, 216
Ellis Island, 20, 267n11

Fekete, John, 66, 68–69
Fisher, Fay Andrew, 165, 183, 196–97, 202, 233
Fiume, 15, 17, 22, 266n4
Fulop, Ilona, 125, 152, 158–59, 163, 166, 181, 183, 196, 212, 275n7, 275n8
Fulop, Mr. (brother to Ilona), 124–25, 275n6

Gary, West Virginia, 93–96, 151
Gleason, West Virginia, 65, 68, 107
Guffey Act, 174

Hajnal, Louis, 159
Harkel, Joseph, 159
Himler, Géza, 1–2, 4–7, 9–13, 16, 24, 43, 152, 164
Himler, Jakab, 1, 10–11, 13, 16, 116, 121, 241
Himler, West Virginia, 160–61
Himler Coal Company, **130**, **132–33**, 161–62, 168, 170–72, 174–77, 182–91, 194–97, 199–200, 277n1, 279n2
Himler Coal Company's Board of Directors, 160–61, 180, 189–91
Himler State Bank, 162
Himlerville, Kentucky, 96, **130–32**, **139**, 169–71, 173–74, 176–77, 179–80, 182–92, 194–200, 203–4, 279n2, 281n3
Hock, John Reverend, 179
Hock, Squire, 150–51
Hocking Coal Field of Ohio, 109, 273n2
Holden, Colonel (leader of hoboes), 70–72, 74
Holden, West Virginia, 143, 148–49, 210
Horthy, Miklós, 165–66, 216–17, 226, 234–36, 239–40, 278n7, 278n8, 285n9, 285n11, 289n5, 291n3
Hotel de Gink, 75–76
Hungarian Revolution of Kossuth, 152, 276n3
Huntington, F. R., 177, 191–92, 194, 197, 279n4
Huntington Herald-Dispatch, 193
Huntington National Bank of Columbus, 177, 191, 279n4

Icza (love of Himler's life), 12, 17, 83, 101, 111, 121, 195, 249–50
Imrédy, Béla, 213, 234, 236, 285n9
Iselin, Pennsylvania, 59–62, 68
Istanbul, 219, 221

Jackson, Robert Justice Chief, 229–32, 237, 288n30, 288n31
Jewish Joint, 233–34, 248, 289n4
Jewish Normal School of Budapest, 6, 10, 13, 60

Kállay, Miklós, 240, 292n6
Karolyi, Michael Count, 150–51, 153, 278n8
Kentucky Board of Education, 170, 184
Kentucky Derby, 73–74
Kermit, West Virginia, 161, 169–70, 186–87
Kermit-Warfield Bridge Company, 171; bridge, 162, 170–71
King's Palace in Budapest, 35, 239, 268n1
Kocsel, John, 92–98
Kormendy, John, 107–8
Kossuth, Lajos, 276n3

Kossuth Hall, 63, 69, 82
Kovach (Hungarian Catholic Priest of New York), 84–85, 87–88
Kovach, George Colonel, 242
Ku Klux Klan, 38, 57, 186–87, 193, 206, 269n6
Kun, Béla, 216, 243, 278n7, 278n8, 285n11

Lajtos (Himler's mining buddy from Iselin, PA), 59–61, 113–17, 122–24
Lang, Eugene, 184
Langley, Katherine, 186, 197, 280n1
Langley, W. John Congressman, 186, 280n1
László, Endre, **137**, 235
Lewis, Catherine, 210
Lewis, John L., 207–11, 284n14
Losonc, 11–12, 17, 265n11
Lusitania, 152, 277n4

Magyar Bányászlap (*Hungarian Miners' Journal*): Holden, West Virginia, 143; New York, New York, 145–55, 158–60, 163, 166, 168; Himlerville, Kentucky, 183, 190–92, 196–97; Columbus, Ohio, 198–200; 202; Cleveland, Ohio, 203–4; Detroit, Michigan, 205, 207, 209–10, 217; Los Angeles, California, 293n4
Marosi, Dezso, 28, 45, 48
Martin County Advertiser, 183, 186–87
Martin Himler Weekly, 192, 199, 207, 217
Mátra Hills, 25, 264n2
Mátraverebély: Himler in, 1–2, 4, 7–9, 15–16, 118, 120–23, 165–67, 180, 242; Himler discusses, 25, 57, 60, 111, 116–17, 169, 172, 205, 218, 241; as trade name for *Magyar Bányászlap*, 145; note, 264n2
May, A.J. Congressman, 155

Mayo Clinic, 195–97, 267n12
McAdoo Tunnel, 22, 61, 86, 272n4
Merry Widow, 18, 266n7
Mihailovic, Draza, 217, 286n12
Miklós, Béla, 238–39, 241, 290n10
Milano, 227, 287n23
Miners' Home, 163–64, 177
Molotov, Vyacheslav, 240, 291n4
Mr. Morgan's Steel Trust, 58, 271n2
"My Old Kentucky Home," 172

Nadassy, Nicholas, 203–5, 218
Nation, 203, 282n5, 282n6
New American, 204
New Boston, Ohio, 36–38
New Camp, Kentucky, 199
New Deal, 205, 207, 271n7, 283n8, 283n10, 284n13
New Republic, 203, 282n6
Norddeutscher Lloyd, 178, 280n5
Norfolk & Western Railway, 30, 151, 159, 161–62, 172–73, 268n7, 276n2
Norris, George, 203, 282n2
Norton, Virginia, 110
Nuremberg, 229–31, 236–37, 288n30, 288n31, 288n32, 288n33

Office of Strategic Services (OSS): Himler discusses, 60, **135**; training, 217, 219, 221; service, 224, 226, 245–46, 249; notes, 271n5, 289n2
Orthodox Jew, 5, 8–9, 264n3

Pásztó, 1–2, 4, 6–9, 11, 56, 118–19, 165, 264n1
Peace Conference of Paris, 247, 249–50, 291n3, 292n2
Petres (landlord of New Boston, OH, boarding house), 38, 40–41, 44–45
Pinson, Ben, 200
Pinson, John, 200

Pittsburgh, Pennsylvania, 30, 57–58, 69, 96, 109, 156, 183, 206
Pocahontas Field, 150–51, 276n2
poppy seed biscuits, 16, 63, 119, 124, 167, 182
Portsmouth, Ohio, 30–32, 34–35, 37–38, 40, 45–46, 48–49, 52, 95, 268n7
Pretoria, 123–24, 275n4

Rakoczi Hall, 22–23, 61, 63, 82
Red Jacket, Michigan, 79–80
Remenyi-Schneller, Lajos, 234
Réthy, Laura, 47
Roebling, New Jersey, 97, 162–63, 172
Roosevelt, Franklin D., 63, 161, 174, 205, 211, 271n5, 271n7, 283n8, 283n10, 283n13, 288n30
Rudnyánsky, Gyula, 47
Ryndam, 113, 274n7

Sabath, J. Adolph Congressman, 205, 283n9
Salzburg, **136**, 226–28, 232–34, 289n2
Savannah, Georgia, 73
Silver Shirters, 38, 269n5, 283n12
Slovak, 9–10, 27, 113, 151, 230, 265n9, 265n11, 278n7
Smallholder Party, 216, 285n10, 289n2
Smith, Gerald L. K., 206, 283n12
Spring Shower, 125, 196, 275n8
SS *Carpathia*, 17, 20–21, 29–30, 32, 45, 50, 61, 63–65, 82, 107, **130**, 266n4
SS *Rotterdam*, 178, 280n6
Staats-Zeitung (German daily newspaper), 64, 99
Sunday, Billy, 127, 276n9
Survey Graphic, 188, 281n2
Sutton, F.G., 191, 194
Szabadság, 56–57, 99, 106

Szabo, Eugene, 18, 78–81, 111
Szabolcsi (Hungarian importer of goods to America), 101–3, 105–9, 112–13
Szálasi, Ferenc, **136**, **137**, 234, 289n5
Szalay, John, 33–34
Szombathely, 120, 122, 179–80, 275n2
Sztójay, Döme Minister Prime, 237, 290n8, 292n6

Teleki, Pál Count, 239–40, 291n3, 292n5
Thacker, West Virginia, 25, 29–30, 35, 68, 267n2
Thacker Mines, West Virginia, 23, 25, 28–30, 35, 38, 49, 59, 267n2, 268n6
Tompkins Cove, New York, 64, 77–79, 81, 111
Transatlantic Trust Company, 156–57
Tsarist Russia, 14, 265n1
Turócszentmárton, 6, 9–10, 265n9

Ulmers (Himler's Pásztó acquaintances in America), 56–57, 62, 69, 87

Vanda, Charles Colonel, 224
Verhovay Segély Egylet, 247, 292n1

Wallace, Henry, 203, 250, 282n3
Warfield, Kentucky, 162, 169–71
Warner, W. H., 199
Weiss, Géza, 16, 61, 104
Welch, West Virginia, 95, 150–51
Wels, Austria, 233
"White Terror" Regime, 167, 216, 179, 278n8
World Federation of Hungarians of Budapest, 213–14, 285n6

Yiddish, 5, 61, 68, 264n4

www.ingramcontent.com/pod-product-compliance
Lightning Source LLC
Chambersburg PA
CBHW060514080526
44586CB00012B/477